OSHA
Compliance
Management

A Guide for Long-Term Health Care Facilities

OSHA
Compliance
Management

A Guide for Long-Term Health Care Facilities

Elsie Tai
g.c.g. risk management, inc.
New York, New York

CRC Press
Taylor & Francis Group
Boca Raton London New York

CRC Press is an imprint of the
Taylor & Francis Group, an **informa** business

CRC Press
Taylor & Francis Group
6000 Broken Sound Parkway NW, Suite 300
Boca Raton, FL 33487-2742

First issued in paperback 2019

© 2001 by Taylor & Francis Group, LLC
CRC Press is an imprint of Taylor & Francis Group, an Informa business

No claim to original U.S. Government works

ISBN-13: 978-1-56670-418-2 (hbk)
ISBN-13: 978-0-367-39776-0 (pbk)

Library of Congress Card Number 00-048127

Library of Congress Cataloging-in-Publication Data

Tai, Elsie
    OSHA compliance management: a guide for long-term health car facilities / by Elsie Tai.
        p. cm.
    Includes index.
    ISBN 1-56670-418-9 (alk. paper)
            1. Industrial hygiene--Handbooks, manuals, etc. 2. Long-term care facilities--Evaluation--handbooks, manuals, etc. 3. Industrial Safety--Handbooks, manuals, etc. 4. Long-term care of the sick--Quality control--Handbooks, manuals, etc. I. Title

RC967.T35 2000
362.1′6′021873--dc21

00-048127
CIP

Visit the Taylor & Francis Web site at
http://www.taylorandfrancis.com

and the CRC Press Web site at
http://www.crcpress.com

Preface

This manual was originally developed for a group of about 130 long-term care facilities in New York State under a special workers' compensation insurance program. The author of this book is the health and safety advisor to this group. She developed highly customized guidance materials specifically for the long-term care industry. These were then combined into a comprehensive manual that has been evolving, and in use, for the past 6 years. The manual embodies everything a long-term care facility and its managers ought and need to know about OSHA, and how to manage the facility most effectively. Discussions on political direction and agency organizational culture, which will affect how regulations are interpreted, inspections are conducted, and how one should proceed in contesting citations are included. The manual presents issues in the order, category, and method as OSHA would view them. Each section provides the basic goal of the standard or subject, a summary, and a discussion of any recent developments, followed by applicable sample policies and appropriate guidance materials.

As the health and safety advisor assigned to this group of long-term care facilities, the author became an integral part of many of the facilities' operations. This provided her with unique insights on the resident focus, Department of Health surveys, Joint Commission surveys (for those that participate), the particular styles of bureaucracy the facilities entail, and even practical matters, such as dealing with dementia, and domestic violence of staff spilling over into the workplace. This intimate familiarity with the industry is reflected in the propriety of the material. Everything is presented to the long-term care operator and management team in an accommodating and palatable form. The author believes that any effort invested in OSHA compliance should also pay dividends in real-life safety management and performance, and other types of liability reduction, rather than be an act in vain in the name of blind compliance.

> "Rules and regulations are for the guidance of the wise, and the strict obedience of fools. Quality never comes from mindless obedience, neither will safety."
>
> —B. Russell.

For the first time, this valuable resource is made available to the general public, outside this local New York insurance program. The author trusts that readers may find it most helpful, effective, and adaptive in managing for safety in the context of a long-term care setting, while meeting compliance initiatives. All the sample policies and guides are available as Word and Excel files on the CRC Web site at http://www.crcpress.com.

This information does not represent legal advice or opinion of current codes. Readers are urged to seek advice of counsel on these issues.

The Author

Elsie Tai has specialized in the environmental and occupational health and safety field for more than 14 years. Her last 9 years have been spent consulting in the long-term care setting within the context of a workers' compensation insurance program aimed at improving facility operational efficacy. Her insights and approach to safety programming and OSHA compliance management uniquely reflect the integration of her continued learning, knowledge, and experience of the issues, culture, and systems particular to this industry. She holds a B.S. degree in environmental science/occupational health and industrial hygiene from Rutgers University, N.J., and an M.S. in organizational leadership from Mercy College, N.Y.

Contents

Chapter 5 In-Service Records

Appendices

1 OSHA, The Agency

I. OSHA REFORM 2000

A. INTRODUCTION

There is a new trend in U.S. government to reinvent anything perceived as inefficient. The move is toward more flexible standards allowing employers more freedom to make informed business decisions on compliance. This is coupled with self-monitoring and self-reporting. The measure of compliance is result oriented; the goal is improvement, rather than compliance with specific prescriptions.

The Environmental Protection Agency (EPA), the Joint Commission on Accreditation of Healthcare Organizations (JCAHO), and the Occupational Safety and Health Administration (OSHA) are turning to risk-based enforcement through greater access to facility compliance and performance data, to help set priorities and make information available. Very limited changes have been made in JCAHO Environment of Care (EC) standards, EPA, or OSHA regulations. However, enforcement methods are definitely changing toward a more performance based system. They call for greater employer disclosure and cooperation, but allow for more flexibility while scaling back specific prescriptions. For example, the EPA 1995 environmental audit policy provides incentives to facilities to self-police, disclose, and correct environmental violations in return for a reduced risk of enforcement actions. The JCAHO controversial new sentinel event policy requires hospitals to self-report adverse events and conduct a root-cause analysis to determine cause and corrective actions, threatening public disclosure and placement on "Accreditation Watch" as the penalty for noncompliance.

Specifically, effective January 1, 1999, Joint Commission surveys will grant more leniency to health care facilities on minor life safety code deficiencies if the facilities have a program in place to monitor and correct, for example, fire safety deficiencies. Although a Building Maintenance Program (BMP) is not required, John Fishbeck, associate director of the JCAHO Standards Department, said "if they don't adopt one, we'll cite them for every life safety deficiency we find." If surveyors find that 95% of a list of life safety program components is functioning properly, then the agency would consider the BMP effective. This is reminiscent of the OSHA Program Evaluation Profile (PEP), the now defunct CCP (Cooperative Compliance Program) initiative, and the replacement target inspection program, where adherence to voluntary guidelines that produces more comprehensive performance will buy employers grace on the usual nitpicking stuff.

In November 1998, a new OSHA policy directive formalizing the agency's partnerships with business and labor, entitled "OSHA Strategic Partnerships for Worker Safety & Health (OSP)," was released. The directive provides guidelines for limited or comprehensive partnerships among local-, regional-, or national-level OSHA and various employers, associations, universities, unions, and even government agencies. These partnerships are voluntary and cooperative relationships designed to achieve measurable goals and establish effective and comprehensive safety and health programs. Therefore, these partnerships are open to employers with poor safety and labor law records, unlike the long-standing VPP (Voluntary Protection Programs) or the SHARP (Safety and Health Achievement Recognition Program). The latter two programs are not covered in this new OSP. Unlike participants in the OSP, the VPP and SHARP participants are afforded inspection exemption incentives.

The OSP incentives to employers are outreach information and assistance in implementing health and safety programs, priority consideration for on-site consultative services (the Consultation

OSHA program, which is designed for smaller businesses with no more than 250 employees on site, and no more than 500 employees companywide); focused inspections should the participant appear on a programmed inspection list; and maximum good-faith leniency in cited hazards, as well as penalty reductions.

Some examples of existing OSPs are the ConAgra/UFCW/OSHA partnership's 5-year program to create models of safety and health excellence at nine plants. Lower absenteeism and turnover rates, improved morale, and reduced injury rates have resulted at one of the plants; the Roofers Partnership of greater Chicago area involves the National Roofing Contractors' Association, the United Union of Roofers, Waterproofers, and Allied Workers, CNA Insurance Company, and the National Safety Council. On the other hand, OSHA recently committed a *faux pas* with the National Association of Manufacturers (NAM) in late 1999 in an OSHA press release after a joint meeting in search of common ground to nurture a partnership. NAM felt the meeting was misrepresented on the press release and that potential good results were lost. NAM accused OSHA of bad manners and demanded an apology.

B.   Target Inspection Program and the PEP—Enforcement Policies

The CCP, the PEP, and other assorted pilot programs continue to mirror this trend in an attempt to meet the new employer and public mandate. The CCP began in 1993 with the Maine 200 pilot program. Maine OSHA sent letters of inquiry to the 200 employers with the worst workers' compensation experience modification in the state. It told them of their status and asked for cooperative response on corrective actions. Of the 200, 197 responded; only 3 did not. Maine OSHA inspected only those three. The other employers worked with Maine OSHA to address their obvious health and safety problems. One of the leading industries with high experience modifications was the skiing industry. In introducing innovative preemployment assessments and other prevention techniques particular to that industry, significant savings were achieved in employers' workers' compensation costs in a relatively short period of time.

A more optimal way for OSHA to identify problem employers would be to utilize experience modifiers and focus on the low end of the spectrum. However, experience modifications are given to oversensitivity toward low or shrinking payrolls, and can appear normal for an industry with a high average rate. Each state-run OSHA office uses a different system to determine target listing. The author believes that the Maine 200 is one of the sounder methods as it targets the very worst performers, using a universal and reliable measurement tool, the experience modifier. Missouri had some serious trouble with the start of its CCP program because it focused on employers with the largest number of injuries, without taking into account the size of the employer and its number of employees. This may have been misconstrued or misunderstood. Nonetheless, with the dismissal of the CCP process, OSHA defaults and resorts to its usual method of prioritizing inspections via the target inspection program. Parameters are similar but slightly shifted. Instead of focusing purely on incident or loss workday rates (LDWII) as a flag, such markers are used in conjunction within the context of specific industries. In the case of long-term care, the probability of inspection has more likely been increased by the dismissal of the CCP. The CCP would have targeted all industries with high rates. Target inspection programs traditionally only target employers with high rates *within* specific industries; i.e., long-term care and construction, thus narrowing the pool of potential candidates. In either case, to the degree that the system could be corrected or was accepted, the selection process is uniform for federal OSHA target inspection programs, but not for state-run OSHA programs.

The federal OSHA offices use a fairly moderate selection method. It is based on randomly selected self-reported surveys from the OSHA 200 Log data on LDWII. The weakness in this type of system is that it begins with a random component. Thus, if one were an employer with a poor incident rate, but were never selected to respond randomly, one could escape notice. Second, it relies on self-reported data, which could be inaccurate. Overreporting may inadvertently include an employer in the Target Inspection List (or CCP or Primary Inspection List in the recent past), at least initially.

However, upon review of the data, and after corrections are made, the employer would be dismissed from the hit list. Underreporting, on the other hand, can go unnoticed; although records verification inspections are conducted, they are uncommon. Third, the cutoff was based on LDWII. This would favor the employer with a limited number of very severe cases of loss time over the employer with a larger number of loss work time cases, which may have been very short, 1 to 2 days each.

C. CHAMBER OF COMMERCE INJUNCTION AND THE OSHA INTERIM PLAN

Ironically, the schizophrenic politics has sent conflicting signals to the agency, which nonetheless trudges along. The local chambers of commerce and many business advocates were in alignment with the CCP and other pilot programs. The national Chamber of Commerce was adamantly against it. OSHA had secured the ability to proceed with its interim plan (which would have happened if not for the CCP), when the Chamber of Commerce brought a temporary injunction on the implementation of the CCP. The court ruled that the interim plan was not associated with the CCP. Since the CCP was struck down in April of 1999, the OSHA interim plan has become, by default, its regular program. The basic difference is that OSHA will prioritize inspections based on high incident rates *within an industry* vs. a basic high incident rate regardless of industry rate and that there will be no cooperative provision so that inspections will be conducted in the traditional manner. This actually worsens the chances of inspection of long-term care facilities since there was a target inspection program for this industry before the CCP. With the CCP, the pool of potential participants was expanded to include any and all industries with a high incident rate vs. a focus on industries with identified high incident rates, i.e., nursing homes and construction. As of the end of 1999, OSHA planned to visit all employers with an incident rate over 16/100. Since the average incident rate for the long-term care industry is 16/100, OSHA planned to visit only 20% of the homes with such a rate, as it cannot possibly visit most of them. Warning letters were sent earlier in the year to affected facilities. This year, the program focuses on all long-term care facilities with an LWDII rate of 14/100 or higher. Letters were also sent again to warn of the high rate and ensuing inspection.

The Chamber of Commerce action is perplexing to many politicians and employer advocates who have been involved in this entire process of OSHA reform. It began with senate hearings on OSHA reform, similar to those the Internal Revenue Service (IRS) has recently undergone. Many different strata of business were represented. The author was involved in providing assistance in developing testimony, along with other consultants, including Russ Phillips, to presenters. Dan Richardson of Latta Road Nursing Home, the Chairman of the New York State Nursing Home Safety Group serviced by the author, was asked to present testimony at these hearings to the Small Business Panel. He had had a harrowing experience with a traditional OSHA inspection shortly before. Like many of the other speakers, his testimony recommended that OSHA become more flexible, business friendly, and less bureaucratic and nitpicking.

D. SENATE HEARINGS ON OSHA REFORM

Recommendations included scaling back the OSHA fixation on specific technical requirements that tend to be especially impractical in long-term care. The bulk of the requirements were created for the manufacturing setting, conducting evaluations based on the quality of the whole system and ultimate results instead of penalizing inconsequential violations, interfacing with relevant industry agencies, i.e., Department of Health (DOH) and JCAHO, focusing inspections on the real problems, and providing assistance to the majority of employers who are well-intentioned and who make reasonable attempts, or at least to leave them alone. The author also pointed out that, although policy changes may be enacted, it would be interesting to see if the actual area offices and individual inspectors, who have been so thoroughly programmed to do otherwise, can effectively change their approach when mandated.

The CCP was the agency's response to the congressional mandate to achieve just this. Besides the cooperative aspect of the CCP, there is also a parallel project of OSHA reform, the PEP. This is a comprehensive assessment tool that evaluates and scores the quality of a health and safety program. The idea was that if one managed a passing grade, a gentlemen's C, the agency would forgo a traditional inspection and focus only on the relevant issues that were flagged. However, if the facility scored less than a 3, equivalent to a grade of C, then the agency would conduct a traditional inspection and issue citations as usual. This was the OSHA method of complying with the congressional mandate without the long and arduous process of passing new legislation or seriously revamping the old. Like the other changing agencies both government and private, as mentioned before, it changed its implementation policy. Nonetheless, this procedure was struck down by the U.S. Court of Appeals for the District of Columbia Circuit. The ruling was that the CCP should have been subjected to public notice and comment. Such a ruling may have an impact on future rule making, in particular, the Safety and Health Program Rule. Meanwhile, OSHA is trying to streamline the rules process, which is often a long and complicated task. It is trying to include private organizations and associations to help set consensus standards. Alas, little has actually changed in the ambiguous OSHA efforts to form "partnerships" with business in light of the schizophrenic reaction of the business sector.

E.   THE CCP—CHAMBER OF COMMERCE INJUNCTION AND FUTURE DEVELOPMENTS

It is this very enforcement policy that was at the heart of the Chamber of Commerce injunction. The chamber argued that the CCP forces employers to comply with standards that had not been properly processed through established regulatory development procedures. The chamber *never* mentions the PEP itself. But it is this enforcement policy that it is referring to. In essence, it is a misunderstanding that the PEP is a standard to which employers are held. Employers cannot be held to any management elements of the PEP, as the elements are *not* standards. However, the PEP *is* used to discern when inspectors should essentially grant leniency and work in a cooperative manner in light of a good health and safety system, and when inspectors should go about their regular business with the traditional fervor and unrelenting fixation on technical details, because the employer does not have a good safety management system. The latter, by the way, is the only method in which OSHA can penalize employers in any case, based on existing standards. There are no penalties, per se, for poor management practices. (This is mirrored by the new JCAHO 1999 policy to grant leniency on life safety deficiencies to facilities that have a building maintenance program, although that is not required. See Section I.A.)

Ironically, it is the CCP pilot program, and not the PEP program, that the chamber attacked. The opposition to the CCP ties in with other arguments including unfair targeting via submitted incident rates, which are interpreted as self-incriminating. The agency was asked, by congressional mandate to focus its energies and attention on employers with problems, and leave good performers be. OSHA inspection plans had earlier been more randomized and less focused. The CCP enriches their existing target industry initiatives by expanding its focus on all industries and employers with above-average incident rates, regardless of their industry category.

The crux of the Chamber of Commerce argument was based on the interpretation that, even though the PEP itself had not been through the usual standards promulgation procedures, yet employers would be held to its standard. This is not necessarily the soundest interpretation since no employer would ever be cited for any deviations from the PEP. It is an assessment tool and, at times, merely a minimum standards screening tool. A good assessment allows the inspector to be employer-friendly. The inspector would only address real issues and perform a "focused inspection" on legitimate items that arose during the assessment. If the result of the PEP is not a green but a red alert, then the inspector would not have the alternative to truly address the PEP issues that were identified, since the PEP is not a standard. Instead, inspectors would proceed with a traditional inspection and cite for every real and inane violation that can be found, as per usual. With the death of the CCP, a myriad of other pilot programs, including the extra discounts for good faith and the PEP, are sporadically applied and experimented with by individual inspectors.

Employers are often concerned about the requirement for employee involvement in these health and safety program standards or program evaluations. Upon reviewing the PEP and this draft Safety and Health Program Rule proposal, it would be evident that the standard required is not invasive but rather common sense, and typical of well-run programs. For example, there should be a system of communicating with employees; employees should have a system to report safety hazards, violations, complaints, suggestions, and the system should function so that these are addressed; employees should be involved in self-inspections. These are not radical elements. They are prudent and efficient aspects of a well-run safety and health program.

Meanwhile, a draft proposed Safety and Health Program Rule has been released by OSHA. A rule is different from a standard in that it is not technically or hazard specific, or prescriptive, and, instead, is a performance-based administrative rule, much as the record-keeping rules are. Much of the content of this rule is, again, reflective of the substance of the PEP. For example, the health-care employee advocates are also pushing for a tuberculosis standard, a new rule revising the OSHA 200 Log Recordkeeping requirements has been in the works for almost a decade and is rumored to be completed in the "near future."

The controversy over ergonomics will continue to heat up in the future. Congress had tied the OSHA budget to the condition that it not release an ergonomic rule in the past few years. Draft ergonomic rules have been circulating. Now, Congress and others less enthusiastic about the rule are asking for OSHA to delay this rule until the National Academy of Science research on ergonomics is completed. OSHA and proponents of an ergonomic standard say that the research project is another delay tactic. They also assert that the facts are viable enough to proceed with a standard and do not require further legitimization by such a research effort. The draft ergonomic standard published in November 1999 was rife with controversy ranging from benefits determinations overriding existing state workers' compensation systems, to job preservation requirements that extend beyond the existing Family Medical Leave Act (FMLA) directives, to triggers mandating implementation of an ergonomic program. The author submitted testimony to the public hearings held in early 2000 addressing some of these concerns.

The rather belligerent reactions of the national Chamber of Commerce may confuse issues temporarily. However, this employer-staged movement will probably continue, despite this blip, as evidenced by the new directives on OSHA Strategic Partnerships (OSP). Those politicians, employers, and employer advocacy organizations involved in OSHA reform are not only perplexed by this legal maneuver, but some are concerned that it would only serve the unions' agenda to return to the traditional system, which hints of harassment more than comprehension. This sentiment was specifically noted by ORC (Organizational Resource Counselors) a Washington-based employer advocacy think tank funded by the top-50 *Fortune* companies.

Ford Motor Company filed a motion asking for leave to intervene in this case on the OSHA side, on December 1, 1998. A spokesperson for Ford said that the company believed that the CCP provided eligible employers such as Ford with an excellent voluntary chance to work cooperatively with the agency and their own employees to improve safety performance. Ford asked the CCP stay for Ford be lifted to allow it and the agency to go forward. Ford admitted that it had not understood the scope of the initial OSHA directive, but now has begun to appreciate the consequences of the stay. Ironically, the American Health Care Association intervened on behalf of the challengers at the same time.

F. THE INTERIM PLAN AND NEW INSPECTION STRATEGY

OSHA has returned to its original inspection plan, which is more industry specific. The plan targets employers in "high hazard industries" (industries with above the average incident rates for general industry), i.e., long-term care, and not only any and all employers with high incident rates in all industries. The cooperative piece is obviously not available, although it has yet to be seen how that would have played out. The pilot program for the new inspection strategy using the PEP began over 2 years ago along with a host of other pilot programs designed for flexibility, customization, and responsiveness to the level of quality of the safety program and commitment of the employer. Although New York State was not one of the six states that participated in the pilot programs, it has been utilizing it in inspections. It appears that this will continue, regardless of the ill-fated CCP program, which has more to do with relationships between employers and the agency, and how inspections are selected. Once an inspection begins, inspectors can continue to experiment with utilizing the PEP and other pilot programs.

Unlike much of the highly technical and jargon-plagued materials generated by the agency, the *PEP is actually a high-quality tool,* well crafted and user-friendly. Long-term care facilities *should* use it in-house as a benchmarking tool, to increase safety performance, control losses, and increase operational leverage. These are the main reasons to reflect on the PEP and attempt to maintain or improve one's scores. It is an added benefit to employers that a passing score on the PEP basically provides the employer with an easier, more cooperative inspection. Poor scores on the PEP are a flag that the facility has more serious problems than OSHA compliance alone. Meanwhile, failing scores would trigger a typical inspection, which would have occurred anyway, even if there were no pilot programs.

G. OSHA Update—At a Glance

1970 OSHA formed during the Nixon administration (serious union considerations)
- Random inspections
- Special emphasis programs on high-hazard industries (construction, etc.)

1992 Senate hearings on OSHA reform
- Found that traditional OSHA nitpicks, issues irrelevant citations, misses the big picture
- Mandated more a cooperative, business-friendly and *systems* approach

1993 CCP—Cooperative Compliance Program (state programs)
- Different selection criteria and strategies, but the same idea—focus on bad actors based on statistics (workers' compensation or BLS/OSHA 200 Log surveys, numbers of claims, etc.) given choice to clean up cooperatively with the application of new pilot programs, or be cited the traditional inspection way.
- One of several pilot programs in formation

1994 Special emphasis program formation for healthcare
1995 Special emphasis program formation for nursing homes
1996 Special emphasis or target program of nursing homes is launched on October 7, 1996.

1997 CCP implemented along with other state pilot programs by federal OSHA
- Focus on *all* industries above national injury rate based on OSHA 200 Log
- Other ongoing state pilot programs to be used in conjunction with CCP
  Program Evaluation Profile—New inspection strategy (systems safety and evaluation of management system over specific physical findings)
  Pilot Penalty Scale—Higher "good faith discounts, better discounts, and lighter penalties for smaller businesses with good faith and larger, more punitive penalties for larger businesses that are 'willful' or egregious."

1997 Chamber of Commerce reaction
- Misunderstood origins, intent, implications, process, and consequences

1997 Temporary court injunction of CCP

1998 Resume interim plan: special emphasis program
 * Records inspections and special emphasis programmed inspections of nursing homes are now *ongoing probably at a higher rate than under the CCP*

1999 CCP struck down: special emphasis program resumed
 * 20% of long-term care facilities with incident rate of 16/100 or higher shall be visited, letters of notice disseminated
 Program Evaluation Profile—New inspection strategy (systems safety and evaluates management system over specific physical findings)
 Pilot Penalty Scale—Higher "good faith discounts, better discounts, and lighter penalties for smaller businesses with good faith and larger, more punitive penalties for larger businesses that are 'willful' or egregious."
 * Controversial draft ergonomic standard published (November 1999)
 * The more important but less-known bloodborne compliance directive is issued (also in November 1999) emphasizing the focus on needleless technology systems, sharps containers, disinfection schedules, medical waste handling, etc.

Records inspections and special emphasis programmed inspections of nursing homes are now *ongoing probably at a higher rate than under the CCP.*

II.   CONTACT WITH OSHA AND REPORTING REQUIREMENTS

A.   RESPONSIBILITY TO CONTACT OSHA

It is the employer's responsibility to contact OSHA *within 8 hours* if there is a work-related fatality, or if three or more employees have been hospitalized for a work-related injury or exposure as a result of one incident. This is catastrophic traumatic event reportage. The three or more hospitalization must be caused by one event. Thus, if three CNAs happen to hurt their backs in the same day, lifting different patients at different times, and were all hospitalized, this would not be reported. Although motor vehicle fatalities are the leading cause of workplace fatality, generally, they are not reported, even though they really ought to be; for example, truck drivers, salespeople, or anyone who happened to die in a car accident while performing work duties. A car accident that involved three employees requiring hospitalization should be reported.

At no other time is the employer required to initiate contact with OSHA.

B.   OSHA CONSULTATIVE SERVICE

An outreach program may solicit a facility for the OSHA consultative service, or a facility may decide to contact the OSHA consultation service to arrange for a consultative inspection. This consultative OSHA is a distinctly different program from the enforcement office. Such a consultative inspection would render a facility immune from regular OSHA enforcement programmed inspections for 1 year. It would not be immune to OSHA inspections as a result of a complaint, fatality, or multiple hospitalizations. The consultative service is not supposed to, and does not, relay information to the enforcement office. There is a very rare chance that a facility may disagree and refuse to comply with the consultative OSHA recommendations that it feels very strongly about. In theory, the consultative OSHA is supposed to then notify the enforcement OSHA office. However, this has never occurred in the history of the programs.

If a facility engages in the OSHA consultation program, it greatly lowers its chance of a programmed inspection for that year. Consider also how helpful a mock DOH survey would be, conducted by a different team from the one officially assigned to the facility. The format would be the same; the basic and essential elements would be visited. It would be good practice in the general sense of practice, procedures, and expected standards. The facility may gain some insights, in effect with no fear of reprisal. However, the styles of approach and handling, along with individual pet peeves and sacred cows would obviously vary greatly. Sometimes, it is these qualities that can give rise to the greatest difficulties with a formal inspection.

C. OSHA-INITIATED CONTACTS BY MAIL

The Bureau of Labor Statistics had traditionally sought injury statistics by randomly soliciting employers to submit their OSHA 200 Log data. This task was transferred to OSHA, under the Department of Labor (DOL), 2 years ago. DOL/OSHA randomly mailed surveys to 65,000 to · 80,000 employers with 60 employees or more. It requests totals from their OSHA 200 Log, and *actual hours worked*. This is reflective of the trend toward self-disclosure and self-monitoring in a risk-based performance system. It takes approximately 30 min to complete the survey. There is a 92% response rate to this survey.

If a facility receives this survey, it should ensure that its OSHA 200 Log record keeping is indeed accurate. Overreporting inflates the industry average and makes a facility a target. The data will be reviewed for accuracy. The data collection personnel conduct follow-ups on data they receive to verify and correct entries that appear to be misconstrued. If there is any suspicion of underreporting, it is investigated. The reviewer would question and verify that the proper definitions have been applied to cases in determining if they are reportable or not, and how they should be recorded.

OSHA now addresses *informal* complaints by mail. In an effort to streamline operations, and maximize resources, OSHA is now investigating *all* complaints, and investigating informal complaints via fax and mail. If the complaint is informal, either anonymous or unwritten, or both, then the agency will send a letter to the employer notifying it that the agency received a complaint, the nature of the complaint, and requesting a response. These inquiries can usually be resolved via phone, fax, and mail without on-site inspector visit.

D. OSHA-INITIATED CONTACTS ON SITE

OSHA will conduct an on-site investigation in response to formal complaints that are written and signed. This is usually done without warning. These issues can be resolved via phone, fax, and mail, if the issue is decidedly not serious, or is easily corrected. Otherwise, the inspector may or may not return to verify that corrections are being made, or that the agreed-upon corrections have been made appropriately.

Programmed inspections are usually scheduled without warning. The Occupational Safety and Health Act specifically mandates that the employer not be given notice of an ensuing inspection. OSHA will, of course, arrive unannounced should there be a fatality or a traumatic incident causing multiple hospitalizations. It will usually appear during normal business hours and workdays. It is highly uncommon for an inspector to appear off-hours, nights, or weekends unless it is deemed necessary by the nature of the complaint or traumatic event. In fact, many long-term care facilities that have experienced OSHA inspections found that the inspectors kept "banker's hours."

E. GENERAL RESPONSE OF A FACILITY

In any of the above instances, one should verify the nature and scope of the inquiry and minimize responses to only the affected issue(s). Do not allow the inspector to veer off and/or beyond the stated and intended scope of the inspection that is established during the opening conference. Do not admit noncompliance. Be courteous, friendly, and respectful. *It is very important to appear cooperative.* Understand that the inspector is not only responding to a complaint, but is trying to discern if the complaint is truly valid, or if the agency is being used as a pawn by a disgruntled employee. In programmed inspections, establishing a good rapport with the inspector will make the rest of the process proceed more smoothly. One will also be able to garner as many discounts and credits as possible for good faith. Thus, it is most beneficial to appear open, straightforward, and helpful, while remaining careful not to incriminate oneself or divulge information that was not specifically sought.

F. RECEIVING AN INSPECTOR

Examine the inspector's credentials. There have been instances where salespeople, union activists, and others have misrepresented themselves as government compliance officers. Have the inspector sign in. Log all the interactions and activities of the facility with the inspector and agency from beginning to end, for the protection of the facility (telephone calls, interviews, requests for documents, content and nature of review, etc.).

Ensure that the inspector is always accompanied by a designated staff person. This individual should be equipped with a pen, clipboard, notepad, tape measure, film, camera, and keys. The person should take notes in parallel with the inspector. Photographs should be taken of issues in question. Always photograph whatever the inspector photographs. (OSHA inspectors are now more cognizant of the need to secure written permission from patients to have their pictures taken.) The staff person should be well acquainted with the operations and the layout in order that he or she can answer questions appropriately, and ask questions of the inspector, while making the case in defense of the facility whenever possible.

This person should also have ready access to all areas, including boiler and utility rooms, storage, all departments, etc. to expedite the process of inspection. Detaining inspectors to find keys or staff with keys only increases the opportunities for inspectors to speak to more people, look longer and harder at things, see something they missed, or dwell on something that they may otherwise have chosen to disregard.

At the closing conference, try to find out what issues concern the inspector. Address the issues on the spot as much as possible. Ensure that notes are complete and reflect the inspector's notes. Sometimes inspectors may attempt to slip something in later that was not mentioned during the closing conference. They cannot embellish upon their findings beyond what was reported and discussed at the closing conference. Look to produce evidence of good safety management in evaluating hazards, developing abatement methods, and following up on it to verify effectiveness.

G. PREEMPTIVE STRATEGY AND THE WARRANT ADVICE

Based on the advice from many seminars conducted by law firms, clients of labor lawyers are often told to ban strictly OSHA compliance officers from their premises until the officers obtain a proper warrant. Some clients are even furnished with elaborate corporate policies and notices to post at the entrance. This is very ill-advised. While employers certainly have a right, by the written word of law, *not* to permit compliance officers access onto their site, in reality, this is taboo.

Attorneys and any other experts who attempt to extend their expertise into an area unfamiliar to them can obtain copies of the Occupational Safety and Health Act and OSHA standards. However, interpretation of these materials without meaningful real-life experience and successful interactions with the agency tends to result in this kind of academic advice, which could wreak practical havoc. This is not unlike handling of OSHA citations (see Section II.I).

Restricting compliance officers from a site and forcing them to get a warrant guarantees loss of any good faith penalty adjustments, which might otherwise have been available, that would have afforded leniency and discounts on penalties. Today, adjustments can be worth up to an 80% reduction. Meanwhile, the tactic ensures the wrath of a harried inspector. Besides, it only takes them 24 hours to obtain an administrative search warrant and return. There is little that one can do in such little time that is worth the liabilities such an action generally incurs.

H. PERSUASIVE STATISTICS

Consider that based on public data on inspections, OSHA cited on average 2.6 citations to cooperative employers and 4.6 violations to those that initially denied entrance on site. In addition, the proposed penalties for the uncooperative employers were approximately 73% higher. In 40% of the cases where employers allowed immediate inspections, OSHA did not cite any violations. But for companies that denied entry, the agency spared only 1%. There was, however, almost no difference in the penalty amounts and rates for serious violations.

I. CONTEST STRATEGY

Upon receiving a citation, it would be best to contact the inspector and possibly his or her supervisor to negotiate the issues in an attempt to lower the fines, to lower the severity of the citation, and/or to dismiss the citations altogether, as quickly as possible at minimal cost and grief. This is better done through technical dueling than by legal means. There is less leeway in the legal aspects of OSHA compliance. Legal action also diverts the focus of attention from what the agency is most concerned with, the technical compliance of the site.

There is much greater advantage and likelihood of winning on technical grounds, rather than on legal grounds. Technical arguments should be grounded in the view that (1) the hazard does not exist; (2) the hazard exists but is not as severe as the inspector has determined it to be; and, last, (3) the hazard had existed, but is abated in an effective fashion, which may not conform to OSHA expectations, but the hazard is indeed abated, nonetheless.

There are more opportunities and flexibility in satisfying the inspector's technical concern and remedying the situation at the lowest level of government as possible. There is no need to inflate the situation by going around the technocrats straight to the bureaucrats. This annoys the technocrats, who had the ultimate authority to grant technical approval of corrective actions or responses. It is best to bring strong evidence to the inspector's supervisor, explaining that the individual doing the inspection was wrong, or made a mistake. The district supervisors are authorized to dismiss violations or to downgrade or upgrade the severity of violations.

Once a case reaches the solicitor's office, most of the aspects of the citations are cast in stone, and the technocrats who crafted it are now out of the picture. The only recourse available in defense is now a matter of law, and not compromise and negotiation. The tone of the dealings with the agency is also then established to be adversarial. Meanwhile, the facility will be paying a premium for its attorney to join it in the quagmire.

III. CITATIONS, PENALTIES, AND CONTESTS

A. INTRODUCTION

Along with the CCP pilot programs and the new strategy of inspection using the PEP pilot program, OSHA has also embarked on a pilot program of penalty reforms. The policy is revised to be more forgiving toward smaller and more responsive employers and tougher with the rest. These pilot programs are experimental by definition. This is the stage for the agency to work out the kinks and to determine if the program is feasible.

B. THE (NEW) PENALTY AND CITATION CEILINGS

The maximum fine for a *willful* or *repeated* violation of an OSHA safety and health standard was raised from $10,000 to $70,000 several years ago as a result of legislation passed in late October 1990 as part of Congress's fiscal 1991 budget package. The minimum for a willful violation was set at $5000. The budget deal, signed by President Bush, also increased the maximum fine for serious violations from $1000 to $7000.

TABLE 1.1
Basic Penalty Structure

Nature of Violation	Old Maximum	New Maximum
Nonserious (see penalty revision section)	$1,000 ($0 in pilot program)	$7,000
Serious	$1,000	$7,000
Failure to abate	$1,000	$7,000
Repeated	$10,000/day	$70,000/day
Willful	$10,000	$5,000–$70,000

Initially, it was feared that higher penalties, and the perception that OSHA will issue fines to raise revenues as part of the overall deficit reduction plan, would create a more adversarial relationship between employers and the agency. It is estimated that $900 million could be raised during this period as a result of the increased penalties. Monies collected through penalty assessment are not banked with the agency that collected it; instead, they become part of the federal budget or deficit. The basic penalty structure is detailed Table 1.1.

OSHA has issued a field directive to its compliance officers that formally sets criteria for the *egregious* citation. OSHA can issue citations on a violation-by-violation basis rather than grouping similar violations under one proposed penalty, increasing the penalty severalfold. This egregious policy can be applied to violations that are clearly willful when one or more of the following apply:

- There are worker fatalities/worksite catastrophe/large number of serious injuries.
- Violations result in persistently high rates of worker injuries/illnesses.
- Employer intentionally disregards safety and health responsibilities.
- Employer's conduct amounts to clear bad faith.
- Large numbers of violations significantly undermine effectiveness of any existing safety and health program.

In setting this type of proposed penalty, OSHA can establish a separate amount for each employee exposed, each piece of equipment affected, or each source of an air contaminant. *OSHA also has been issuing heavy penalties for obvious record-keeping violations, particularly when it finds that cases are not recorded.* This now means that an OSHA inspection *could* conceivably result in a major OSHA fine.

C. Scope and Probability of Inspection

States with OSHA state plans tend to conduct more frequent, but less comprehensive inspections than federal OSHA does. There is often confusion about the state or federal jurisdiction of the OSHA office. For example, New York State does not have its own state OSHA plan. Although the OSHA in New York State may sometimes be referred to in some manner delineating the geography and jurisdiction of the particular office, New York State OSHA offices are federally administered. New York State is part of the federal OSHA Region II division. New York State, as well as Connecticut, however, have a state plan for public employees only (PESH, Public Employees Safety and Health).

The following states and territories have approved OSHA state plans: Alaska, Arizona, Hawaii, Indiana, Iowa, Kentucky, Maryland, Minnesota, South Carolina, Tennessee, Utah, Vermont, Virginia, the Virgin Islands, and Wyoming. At the present rate of inspections, OSHA would take 50 years to inspect every private company in New York State, as compared with 10 years for Nevada.

D.   TRADITIONAL VS. REVISED PILOT TEST OSHA CIVIL PENALTIES AND ADJUSTMENTS

1.   Size Adjustment (old)

The size adjustment factor is traditionally as follows: For an employer with only 1 to 25 workers, the penalty will be reduced 60%; 26 to 100 workers, the reduction will be 40%; 101 to 250 workers, a 20% reduction; and more than 250 workers, there will be no reduction in the penalty.

2.   Size Adjustment (test)

Pilot test penalty for size adjustments is as follows: For employers with 1 to 10 employees, an 80% reduction; 11 to 30 workers, 65%; 31 to 100 workers, 40%; 101 to 250 workers, 20%. There is no size adjustment for employers with 251 or more employees. If the inspection finds any violations classified as high-gravity serious, willful, repeated, or failure to abate, the size adjustment will be limited to no more than 40%.

3.   Other-Than-Serious Adjustment (test)

The penalty revisions to be tested include no penalties for other-than-serious violations, if the employer has 250 or fewer employees companywide at all times during the past 12 months and the current inspection does not reveal any willful, repeat, or failure-to-abate violations. This policy does not apply to violations of regulatory requirements such as requirements for record keeping or for posting citations.

4.   Good Faith Adjustment (old)

Traditionally, there was up to an additional 25% reduction for evidence that the employer is making a good faith effort to provide good workplace safety and health, and an additional 10% reduction if the employer has not been cited by OSHA for any serious, willful, or repeat violations *in the past 3 years.* To qualify for the full 25% good faith reduction, an employer *must have a written and implemented safety and health program* that includes programs required under the OSHA standards, such as hazard communication, lockout/tagout, or bloodborne pathogen, etc.

5.   Good Faith Adjustment (test)

The test program allows for good faith adjustments up to an 80% reduction based on evaluation of overall safety and health program (score of 3 or better on the PEP). No more than 40% reduction for good faith is given if the worksite has a lost workday injury rate at or above the national average *for the industry,* if there are any high-gravity serious violations, or if there are more than a few total violations. If the worksite qualifies for a 60 to 80% good faith reduction, no penalties shall be proposed for other-than-serious violations.

6.   High-Gravity Serious Violations (test)

The penalty for high-gravity serious violations is increased to a mandatory $7000, the statutory maximum. Employers who had a previous OSHA inspection within the last 5 years with no citation for serious, willful, repeat, or failure to abate violations would receive a penalty reduction of 10% based on history.

E.   SERIOUS VIOLATIONS PENALTY GRADING STRUCTURE

1.   Serious Violations

The typical range of proposed penalties for serious violations before adjustment factors are applied will be $1500 to $5000, although the regional administrator may propose up to $7000 for a serious violation when warranted. A serious violation is defined as one in which there is substantial

---

TABLE 1.2
Penalty Grading Structure: Serious Violations

Severity	Probability	Penalty
High	Greater	$5,000 ($7,000 test mandatory)
Medium	Greater	$3,500
Low	Greater	$2,500
High	Lesser	$2,500
Medium	Lesser	$2,000
Low	Lesser	$1,500

Note: Penalties for serious violations that are classified as high in both severity and greater in probability will only be adjusted for size and history.

---

probability that death or serious physical harm could result, and the employer knew or should have known of the hazard. Serious violations will be categorized in terms of severity as high, medium, or low, with the probability of an injury or illness occurring being greater or lesser. Note above that in the test pilot penalty system high-gravity serious violations carry a mandatory $7000 fine. Base penalties for serious violations will be assessed as detailed on Table 1.2.

2. Other-Than-Serious Violations

If an employer is cited for an other-than-serious violation, which has a low probability of resulting in an injury or illness, there will be *no proposed penalty.* The violation must still be corrected. If the other-than-serious violation has a greater probability of resulting in an injury or illness, then a base penalty of $1000 will be used, to which appropriate adjustment factors will be applied. The OSHA regional administrator may use a base penalty of up to $7000 if circumstances warrant.

3. Regulatory Violations

Regulatory violations involve violations of posting, injury, and illness reporting and record-keeping requirements, and not telling employees about advance notice of an inspection (when OSHA has made previous arrangements for whatever reasons). OSHA will apply adjustments only for the size and history of the establishment.

Base penalties, before adjustments, to be proposed for posting requirement violations

- Annual summary, $1000
- Failure-to-post citations, $3000

Base reporting and record-keeping penalties

- Failure to maintain OSHA 200 and OSHA 101 forms (or in-house accident/incident forms that reflect the identical information as the 101), $1000
- Failure to report a fatality or catastrophe within 8 hours: $5000 (with a provision that the OSHA regional administrator could adjust that up to $7000 in exceptional circumstances)

- Denying access to records, $1000
- Not informing employees in advance of an inspection, $2000

In the past, failure to have the OSHA notice posted was most frequently cited and $1000 fines were assessed. This was the first inane technical nitpick that was done away with. In June 1996, President Clinton issued a memorandum to all OSHA offices demanding that the practice of citing this penalty be stopped immediately. This was the first outcome of the Senate hearings on OSHA reform.

4. Willful Violations

In the case of willful serious violations, the initial proposed penalty must be between $5000 and $70,000. OSHA calculates the penalty for the underlying serious violation, adjusts it for size and history, and multiples it by 7. The multiplier of 7 can be adjusted upward or downward at the OSHA regional administrator's discretion, if circumstances warrant. The minimum willful serious penalty is $5000. Willful violations are those committed with an intentional disregard of, or plain indifference to, the requirements of the Occupation Safety and Health Act and regulations.

5. Repeat Violations

A repeat violation is a violation of any standard, regulation, rule, or order where, upon re-inspection, a *substantially similar* violation is found. (*Note:* Repeat violations can be issued for a facility that has never received a citation if a sister facility under the same management/ownership had received the similar citation previously.) Repeat violations will only be adjusted for size, and the adjusted penalties will then be multiplied by 2, 5, or 10. The multiplier for small employers, 250 employees or fewer, is 2 for the first instance of a repeat violation, and 5 for the second repeat. However, the OSHA regional administrator has the authority to use a multiplication factor of up to 10 on a case involving a repeat violation by a small employer to achieve the necessary deterrent effect. The multiplier for large employers, 250 or more employees, is 5 for the first instance of a repeat violation, and 10 for the second repeat. If the initial violation was other than serious, without a penalty being assessed, then the penalty will be $200 for the first repetition of that violation, $500 for the second repeat, and $1000 for the third repeat.

6. Failure to Abate

Failure to correct a prior violation within the prescribed abatement period could result in a penalty for each day the violation continues beyond the abatement date. In these failure-to-abate cases the daily penalty will be equal to the amount of the initial penalty (up to $7000) with an adjustment for size only. This failure-to-abate penalty may be assessed for a maximum of 30 days by the OSHA area office. In cases of partial abatement of the violation, the OSHA regional administrator has authority to reduce the penalty by 25 to 75%. If the failure to abate is more than 30 days, it may be referred to the OSHA national office in Washington where a determination may be made to assess a daily penalty beyond the initial 30 days.

F. PENALTY REALITY

Historically, 6% of cited employers have contested their penalties. OSHA assessed maximum penalties in only 2.1% of violations that carry fines. Only 14% of the largest companies received penalties over 50% of the maximum.

When OSHA has settled a case with an employer, it has reduced its proposed penalties by an average of 72%. When the Occupational Safety and Health Review Committee (OSHRC) has decided a case, it has reduced proposed penalties by an average of 96%. When OSHA has settled, the agency has downgraded 14% of cited willful violations and withdrawn   %. The OSHRC downgraded 57% and withdrew 36%.

G. CONTEST STRATEGY (SEE SECTION II.I)

IV. STATISTICS AND FORMULAS

A. INTRODUCTION

Injury statistics drive the OSHA agenda. The agency relies on injury statistics to determine the levels of risk experienced in different industries and job categories over time. The agency even attempts to link increases and decreases in rates to certain developments and events, including technology, procedures, demographics, and even the implementation of specific standards and the formation of the agency itself. Injury statistics at individual facilities as viewed from the OSHA 200 Log are used to determine the level of risk at a given facility. It is an essential component in the overall impression of the employer, and the elements that are balanced in deciding penalties.

B. INCIDENT RATES

Injury statistics that are reviewed often take the form of incident rates. Rates are an excellent method of statistical comparison, in that they provide a common denominator to the main variable, the number of employees. Often, employers review their totals of all injuries and categorical injuries. However, these numbers can be misleading, as they do not account for changes in the number of employees, or hours of exposure. Therefore, even as number of injuries drop, the *rate* of injuries may in fact be the same, or even worse if the number of employees or hours worked has decreased. Conversely, even as the number of injuries increase, the rate of injuries may in fact be the same, or improved if the number of employees or hours worked has also increased. A continual point of reference is provided when discussing incidents or injuries or categories of injuries expressed by the number of such cases in terms of a rate, per 100 employees.

An incident rate for any category of injuries can be constructed. Typically, the categories are taken from the bottom totals of the columns of the OSHA Log. For example,

- To find the frequency rate, use the total number of entries in the Log (column A).
- To find the severity rate (LWDII), use the total from column 2 of the Log.

One can even calculate the entire incident rate by using the total number of incidents (whether or not the incident was recordable on the OSHA 200 Log) to find the in-house rate of all reported incidents within a facility. This is a useful benchmark, as well.

Warning: Do not ever use this number, or any of these numbers, generally speaking, as the basis for measuring or rewarding safety performance. Such a system is not so much an incentive to improve reporting of incidents, root-cause analysis, follow-up, communication, teamwork or safety performance, as it is an incentive to underreport injuries and to curb open discussion of near misses, incidents, and hazards. Such a system would also contribute to an inferior grade on the PEP, which recognizes this as counterproductive to managing safety systems.

C. THE INCIDENT RATE FORMULA

Use the total of the type of incident selected above in Equation 11, placing it in the (?) slot to calculate the incident rate of that type of incident category. This formula is based on incident per year as reflected by the 200,000 factor, which represents the hours worked by 100 full-time employees in 1 year.

$$\frac{\text{No. of Cases of (?)} \times 200{,}000}{\text{Actual Hours Worked}} \qquad (11)$$

In calculating actual hours, it is acceptable to use the number of employees multiplied by 2000 (50 weeks at 40 hours a week). However, this would usually greatly inflate the result since long-term care has quite a large proportion of part-time employees. Also, if payroll data are used in calculating actual hours worked, be careful not to mistake double-time or overtime pay as doubling or inflating the actual hours worked. It is most beneficial to have the maximum sum as high as possible for the actual hours worked since it makes up the denominator, resulting in a lower fraction, or incident rate. However, be sure that the number can be substantiated.

D. USE OF INCIDENT RATES

Incident rates are used in two basic ways. One is to benchmark against the industry average. The second is to benchmark against a facility's own performance over time. A rate can be generated by a department and compared with rates of other departments and to itself over time.

Table 1.3 details the incident rates for both general industry and the long-term care industry, for the years 1995, 1996, and 1997 as compiled by the Bureau of Labor Statistics.

Calculation Note: Health and safety organizations and professionals use incident rates to compare performance over the years, to view trends, and to compare risks among different categories of workers or industries. These numbers are not usually derived on a monthly or quarterly basis in most published accounts. It is generally not terribly important to review rates by months or quarters, because little would be reflected unless (1) the employee population is in the thousands and (2) drastic changes were taking place from 1 month to the next (i.e., radical changes in equipment, staff, procedures, logistics, design, etc.). Incident rates are an *indirect* and *downstream indicator* of safety performance. In most cases, month-to-month variations reflect randomness and probability, especially when the data pool is small (less than 1000 employees).

Nonetheless, there are some facilities that would like to calculate rates by month, quarter, or "to date" in the same manner they do for infection rates and in tracking other epidemiological issues. In that case, Equation 11 is not valid. The formula is based on 100 employees working at 40 hours a week for 50 weeks a year, which assumes 2 weeks' vacation. It does not account for sick days, personal days, holidays, etc. Therefore, if the formula is to be used on a monthly or quarterly basis, it must be adjusted.

To calculate monthly rates, replace the 200,000 factor with 16,666.667 (which is man-hours worked per month or 200,000/12). For quarterly, rates use 50,000 (200,000/4). To get the rate to date, multiply the number of the current month by 16,666.667.

The author has discovered that some facilities may have been using the 200,000 factor for monthly or quarterly incident rates, for several years to more than a decade. These statistics may not be correct, but they are not useless either. Since the calculations were based on the same degree of inflation (12 times the man-hours for monthly rates, and 4 times for quarterly), then the rates provide meaningful *relative value*. In scientific terms, they are *accurate*, but not *precise*.

TABLE 1.3
Incident Rates

Incident Type		General Industry			Long-Term Care	
Year	1995	1996	1997	1995	1996	1997
LWDII	3.6	3.4	4.4	8.8	8.3	8.8
Med only	4.4	4.1	3.8	9.4	8.2	7.5
Total	8.1	7.4	7.1	18.2	16.5	16.2

Note: Total incident rate (Total); rate of loss work day cases (LWDII); and rate of no loss time but OSHA 200 Log reportable cases (Med only).

E.   OSHA 200 Log Review

The Log will always be the first document reviewed during an inspection, unless the inspection was brought about by some devastating traumatic event that required immediate intervention or investigation. An incident rate based on the facility data will be determined. Within the original scope of the CCP, upon verification of the incident rate, an inspection may be canceled if the agency finds that the LDWII rate of a facility is lower than the 7.0/100 employee cutoff rate used in the program. Thus, if the facility had made reporting and calculation errors that had inadvertently inflated its rate and placed it in the CCP program in the first place, the facility can be removed by correction.

Please refer to Appendix F on the OSHA 200 Log for a comprehensive overview on reporting definitions, requirements, and protocols of the OSHA 200 Log.

V.   INTERFACE WITH JOINT COMMISSION

Vice President Al Gore awarded OSHA and the Joint Commission the Hammer Award for their efforts in working together. In reality, very little has been accomplished. It is difficult to describe exactly what they have done. It is not collaboration, per se. It is more acquaintanceship. OSHA inspectors are not at all concerned with DOH issues, Joint Commission survey processes, or any issues involving residents. At best, they are merely learning how their own issues can be illuminated by each other's existing management and survey systems. Unfortunately, there are no plans for OSHA to interface with the DOH which conducts surveys for all facilities annually, whereas only facilities that choose to become accredited by the Joint Commission are subject to Joint Commission surveys.

The following are two crossover references:

1. Joint Commission and OSHA crossover notes—Practical comparisons of the similarities and differences in basic goals, approaches, survey process, and postsurvey activities (Table 1.4).
2. Cross-reference table of OSHA issues as related to Joint Commission by departments— This is a list of OSHA issues with specific corresponding Joint Commission sections referenced, listing by departments that are affected, and to what degree (Table 1.5).

Many of the sample management policies for specific and comprehensive issues, implementation, assessment and record-keeping tools, and training materials are all designed in such a way to provide the type of documentation and demonstration of implementation that the Joint Commission also seeks. (Examples: sample Ergonomic Policy, Health and Safety Policy, Workplace Violence Policy, Health and Safety Activities Log, the Mandatory OSHA Training Record-Keeping Cross Reference Tracker, etc.)

Last, at the end of this section is material from a Joint Commission-generated document, entitled "Meeting Joint Commission and OSHA Requirements." Joint Commission surveyors are supposed to provide on-site education and consultation during interviews and observations that relate to OSHA issues. However, they will in no way conduct OSHA inspections, and will not inform OSHA of the conditions at a facility, or share findings with OSHA (see Table 1.6 and Charts 1.1 and 1.2).

TABLE 1.4
Joint Commission and OSHA Crossover Notes

Joint Commission	OSHA
Overall	
Continuous Improvement Process	Safety Management Program
Performance Improvement Program	PEP Program Evaluation Profile
Underlying requirement of both is to	Both review
1. Identify, assess, and evaluate issues	1. Management plans
2. Intervene or abate	2. Evidence of implementation
3. Monitor, followup	3. Employee training documentation
Performance measurement tools for benchmarking and assessing outcomes	Policies, training documentation, evidence of implementation via employee interviews and inspection findings
Goal is patient care	Goal is employee safety and health
Review conditions and behaviors	Review conditions
The Survey Process	
Opening conference to lay groundwork	Opening conference to explain reason and scope
Visit sites	Inspect
Review safety committee minutes among other documents	Does not require safety committee minutes
Formal interview with leaders	Interview with employees only
Interact/interview staff, patients	Interview on employee health and safety
Interview regarding patient care; review patients' medical records	Employee medical records only as per Log
Review track record 1 year preceding current triannual survey but more if needed	Review 1 to 3 years up to 5-year-Log; review 1 to 3 years training records if needed
Feedback sessions throughout process	No feedback sessions at request of compliance officer
Bimonthly safety committee meetings (unless justified by experience and appropriateness)	Not required (quarterly is okay, more better)
Annual nonresident area inspections (bi-annual resident area inspections)	Periodic site inspections (annually and as needed)
Postsurvey	
In defending a deficiency	In defending a (potential) citation
Leadership conference to try to iron out findings of *preliminary report*	Closing conference lays down the line
Later can expand and revise report	No expansion from closing conference (as nothing can be added later on, ensure all have the same notes)
Demonstrate alternative but adequate management of the issue	Prove hazard does not exist, and/or
	Hazard could not have been reasonably recognized, and/or
	Hazard is abated in an alternative but adequate manner
Type 1 Recommendation with time frame	All citations have specific time frame (varies depending on situation)
Challenge 30 to 20 days depending on type	All contests in 15 working days
Public and confidential records	Almost everything is public record except for internal memos

TABLE 1.5
Cross-Reference Table of OSHA Issues as Related to Joint Commission (and Departments within Facility)

Joint Commission	OSHA	Departments							
		ADM	NUR	DTY	PT	MTN	HSK	LDY	REC
Safety Management LD2.8; EC.1.3, 2.1, 2.2, HR.1, 3, 4, HR 4.2; PI.1, 2, 3, 3.3, 4, 5	Health and Safety Program (Safety and Health Programs)	*	√	√	√	√	√	√	√
Information Management LD.2.3. and 2.7 EC.2.5	OSHA 200 Log	*	√	√	√	√	√	√	√
IC.1	Bloodborne Pathogens	*	*	√	*	√	*	*	√
IC.1	Tuberculosis	*	*	√	√	√	√	√	√
(See first entry above)	Lockout/Tagout	*	√	√	√	*	√	√	√
EC.1.7, EC.2.8	Electrical Safety	√	√	√	√	*	√	√	√
EC.1.5, 1.8, 2.4, 2.5, 2.7, 2.13	Hazard Communication	*	√	*	√	*	*	*	√
(See first entry above)	Ergonomics: General	*	*	√	*	√	√	√	√
PI.3.3.2	Ergonomics: Patient Transfers	*	*	—	*	√	—	—	√
EC.1.4	Workplace Violence: General	*	*	√	√	*	√	√	√
EC.4 to 4.4	Workplace Violence: Dementia	*	*	—	√	√	√	—	*
Fire Safety/EC.1.7, 2.6, 2.10	Emergency Plan	*	*	*	*	*	*	*	*
Fire Safety/EC.2.1	Fire Extinguisher Training	√	√	√	√	√	√	√	√
EC.2.1	Training Recordkeeping	*	*	*	*	*	*	*	*
EC.3, 3.1, 3.2	Machine Guarding	√	—	*	—	*	√	√	—
EC.3, 3.1, 3.2	Personal Protective Equipment	√	*	*	—	*	*	*	—
EC.3, 3.1, 3.2	Confined Spaces	*	—	—	—	*	—	—	—
EC.3, 3.1, 3.2	Accident Medical and Exposure	*	—	—	√	√	√	√	—
EC.3, 3.1, 3.2	Walking and Working Surfaces	—	—	*	—	√	*	—	—
Ventilation EC.1.2, 1.3, 1.5, 1.7, 1.8, 1.9, 2, 2.1, 2.2, 2.4, 2.7, 2.8, 2.11, 2.12, 2.13, 2.14 Utilities Mgmt LD.1–4	Preventative Maintenance	√	—	—	—	*	—	—	—
EC2.11	Hazardous Materials and Waste (Hazcom and Bloodborne)	√	*	*	√	*	*	*	√

Key: * very relevant; √ relevant; — less relevant; ADM–administration; DTY–dietary; PT–physical therapy; MTN–maintenance; HSK–housekeeping; LDY–laundry; REC–records.

TABLE 1.6
OSHA Topics Compared with Joint Commission Standards

OSHA Topics	Joint Commission Standards
Voluntary Protection Program (VPP)	EC.1.3, EC.2.1, EC.2.2: Safety Management HR.1, HR.3, HR.4 PI.1, PI.2, PI.3, PI.3.3, PI.4, PI.5
Ventilation	EC.1.2, EC.1.3, EC.1.5, EC.1.7, EC.1.8, EC.1.9, EC.2, EC.2.1, EC.2.2, EC.2.4, EC.2.7, EC.2.8, EC.2.11, EC.2.212, EC.2.13, EC.2.14, EC.2.8 Utilities Management LD.1, LD.2, LD.3, LD.4
Information Management	EC.3: Information Collection and Evaluation System IM Chapter; HR.4, HR.4.2, HR.4.3, HR.5 PI.1, PI.2, PI.3, PI.3.3, PI.4, PI.5
Patient Handling, Lifting and Moving	PI.3.3.2: Risk Management Data in Improving Organization Performance Chapter
Safety and Health Programs	EC.1.3, EC.2.2: Safety Management
Workplace Violence	EC.1.4: Security
Laboratory and Hazard Communication	EC.1.5 and EC.2.4: Hazardous Materials and Waste
Bloodborne, Tuberculosis, and Legionella	IC.3
ETO, H_2CO, and Glutaraldehyde	EC.1.5, EC.1.8, EC.2.7, EC.2.13
OSHA Record Keeping	IC.3
Hazardous Drugs, Reproductive Hazards, and Anesthetic Gases	EC.1.5, EC.1.8, EC.2.7, EC.2.13
Walking and Working Surfaces	EC.3, EC.3.1, EC.3.2
Fire Safety	EC.1.7, EC.2.6, EC.2.10
Electrical Safety	EC.1.7, EC.2.8
Safety and Health Statistics	LD.2.3 and LD.2.7
MSDS and OSHA 200 Log	EC.2.5
Education/Professional Qualifications of Parties Responsible for the Safety and Health Program	LD.2.8, EC.2.1, HR.4, HR.4.2

Notes: All standards are from the *Joint Commission Comprehensive Accreditation Manual for Hospitals.* EC refers to the "Environment of Care" chapter; IM refers to the "Management of Information" chapter; PI refers to the "Improving Organization Performance" chapter; IC refers to the "Surveillance, Prevention, and Control of Infection" chapter.

Source: The Joint Commission 1997 Environment of Care Series, Issue 3, *OSHA & Environment of Care Compatibilities.* For further information on this publication, please call Joint Commission customer service at 630–792–5800 and ask for publication order code PTSM-845.

CHART 1.1
Hospital Accreditation Services Survey Activities Relevant to OSHA Topics

Day 1
All Surveyors
Document Review Session
Leadership Interview

Physician Surveyor	Nurse Surveyor	Administrator Surveyor
Patient Care Setting Visit	Patient Care Setting Visit	Building Tour
Anesthetizing Location Visit	Anesthetizing Location Visit	
	Patient Care Setting Visit	

Day 2
All Surveyors

Physician Surveyor	Nurse Surveyor	Administrator Surveyor
Pathology and Clinical Lab	Patient Care Setting Visit	Patient Care Setting Visit
Services Visit	Infection Control Interview	Review of Environment Care Documents
Patient Care Setting Visit	Human Resources Interview	Chief Executive Officer/Strategic
		Planning and Resource
		Allocation Interview
		Pharmacy Services Visit
		Human Resources Interview

Day 3
All Surveyors

Physician Surveyor	Nurse Surveyor	Administrator Surveyor
Performance Improvement	Performance Improvement	
Team Interview	Team Interview	

Note: Surveyors will provide on-site education and consultation during interviews and observation activities relating to OSHA requirements. They will not inspect, nor report any information to OSHA.

Source: From *Meeting Joint Commission and OSHA Requirements,* a Joint Commission publication.

CHART 1.2
Description of Example Survey Activities and Relevance to OSHA Topics

Building Tour

Purpose Address issues related to the Life Safety Code in the structure of the building, and performance of organization programs related to patient rights, infection control, and safety management.

Issues to be addressed Conformance of the building with the Life Safety Code, issues related to the life safety management program, and standards in other chapters of the Comprehensive Accreditation Manual for Hospitals (CAMH).

Relevance to OSHA Workplace hazards related to maintaining a safe environment for patients and workers can be identified.

Review of Environment of Care Documents

Purpose To assess through documentation aspects of the implementation of the program of the organization to address issues relating to the management of the environment of patient care, including the plans of the organization for improvement relating to life safety.

Issues to be addressed Compliance with standards in the Management of the Environment of Care chapter of the CAMH.

Relevance to OSHA Tracking and trending of improvements in the environment of care/environment of work are identified.

Human Resources Interview

Purpose Assessment activity focuses on three human resource components: staff orientation, training and education; provision of adequate staff to meet patient needs; and a review of the facility's competence assessment process.

Issues to be addressed Human resources function standards, which include assessment of staff education, the provision of adequate staff, and the facility's competence assessment process. The surveyors may also touch upon Leadership, Management of the Environment of Care, and Improving Organization Performance standards relating to staffing issues.

Relevance to OSHA A review is conducted on the management process for the employee health portion of the safety and health program.

Infection Control Interview

Purpose Assess the infection control program of the organization.

Issues to be addressed Standards in the Surveillance, Prevention and Control of Infection, Improving Organization Performance, and Management of Information chapters of the CAMH.

Relevance to OSHA Bloodborne Pathogens and immunization records for employees.

Source: From *Meeting Joint Commission & OSHA Requirements,* a Joint Commission publication.

2 Inspection Preparedness

I. INTRODUCTION

Inspection preparedness is more about being able to manage an inspection to the best of a facility's ability, than it is about being in compliance with everything. The latter can be exhausting. OSHA inspection preparedness involves the orchestration of documents and interactions in such a way to maximize a facility's strengths and the likelihood of a positive outcome from the experience, and to minimize potential liabilities and interruptions to the operation. It may be enough to *survive* an inspection; it is the author's intent to help readers *manage* an inspection to their advantage as much as possible.

This chapter also provides the tools to guide an inspection process, and prepares the reader for those management aspects that are not as specific as those dictated by standards, as in Chapter 3. This chapter provides the tools to produce materials and exchanges that accentuate any and all safety-related activities, particularly those inherent in a system that may be less obvious, and are often taken for granted, but for which great credit could be taken. This is the *safety management* chapter of this manual. Naturally, in this sense, the Program Evaluation Profile (PEP) is included in this chapter. PEP is the nonstandard safety program management assessment tool by which inspectors can gauge which inspection approach to use, "focused inspection," reviewing only issues of legitimate concern at the time, or "traditional," the whole ball of wax.

This chapter also includes OSHA Inspection Policy and Procedures Guide; Model Sample Health and Safety Program; and Health and Safety Activities Log, a quality assurance or QA-type charting of safety activities, and, more importantly, the explanations on how best to use the Log.

A copy of this sample document, titled "InspectionPolicy.doc," is available on the Web site.

II. GUIDE: OSHA INSPECTION POLICY AND PROCEDURES

This guide should be used to plan and design a facility's policy and strategy on how to handle OSHA inspections.

A. GENERAL POLICY

All employees coming into initial contact with the inspector, and assigned to accompanying the inspector, shall treat the inspector with professional courtesy at all times. Requested information that lies within the scope of the inspection is furnished.

B. RECEPTIONIST

Upon the arrival of the inspector, the receptionist shall be pleasant, and courteously ask for official identification. OSHA inspectors carry badges, which should be verified. The receptionist should have the inspector sign in, and be seated to wait while the appropriate liaison is contacted to meet with the inspector.

C. THE OSHA LIAISON

There should be a hierarchy of three potential OSHA liaisons in the facility. The liaison can be the assistant administrator, the maintenance director, or administrator, or anyone else who may be qualified. More importantly, the liaison should be chosen based on availability and *qualification*. The OSHA liaison should have:

A good knowledge of overall operations of the entire facility

A good knowledge of the layout of the facility and *ready access* to all areas

A command of OSHA *jargon* and *requirements*

The ability to conduct him or herself in a continuously *amenable* fashion

The ability to discern what requested information lies within the scope of the inspection parameter and respectfully to decline that which does not

The ability to entertain legitimate requests for information without inadvertently volunteering additional information and/or enlarging the scope of inspection by such unwitting actions and conversations

D.   VERIFICATION OF INSPECTION TYPE

The OSHA liaison should first introduce him or herself and inquire about the type of inspection that is to be conducted. This is the "Opening Conference" whereby the OSHA inspector is required to provide the reason for the agency's appearance, be it a written complaint that the inspector shall show you, or a programmed inspection, etc. This will determine the scope of the inspection, which should be established at the very start.

E.   KEEPING A LOG

Maintain a running log of all the facility's interactions with the agency, and its representatives. Make note of dates of any and all transactions including telephone calls, receipt of mail, site visits, interviews, the facility's responses in kind, etc. Record the names and titles of OSHA representatives, the people they spoke with on the staff, the content of their interview, the result of their interviews. Follow-up with staff and department heads after each interview to get their side of the story of what was asked, and how they responded. This is especially important if you find the inspector to be difficult or the inspection process is becoming troublesome or unreasonable.

If you find the inspector's behavior or conduct to be unprofessional, unfair, or difficult, it is best to document these as objectively as possible, with quotes, and clear unarguable descriptions of the behavior on the Log (as opposed to denigrating comments on the individual, focus on the inappropriate *conduct*). You may consider complaining to the inspector's supervisor, but that may worsen the situation, rather than improve it. Logging inappropriate and unprofessional behavior may help in the future in having a citation dismissed because of improper investigatory methods used. It would also provide you with a more concrete record of descriptions of problems should you take your case to your Congressional Representative. For better and worse, OSHA inspectors and their respective home offices are rather skittish and fearful of political heat. If you experience an unusually unreasonable inspection, contacting your Representative with strong legitimate issues can bring some relief. However, the same type of political pressure and threats, or the mere fear of it, has also led OSHA to make sweetheart deals with very poor actors who do not deserve such leniency, or to stay clear of a large-profile employer with serious health and safety issues.

F.   LAST-MINUTE OSHA 200 LOG UPDATE

You probably only have a few minutes, but maybe up to a few days, before the inspection begins, after you have been notified. Take advantage of whatever time you have, to immediately check and update your OSHA 200 Log, which will be the first document that the inspector will review. Take care that all incidents that are deemed recordable by the OSHA definitions are, indeed, on the Log. Any PPD (a tuberculosis test) conversions and "exposure incidents" should have follow-up records available on site. Calculate your rate, as they will. Your 1997, 1998, and 1999 rates are very important. They are supposed to focus on the 1998 rates, and alleviate some pressure if your 1999 rates reflect an improvement. (See Appendix F for a review of the OSHA 200 Log.)

G.   LAST-MINUTE POLICY DRAWING (SEE CHECKLIST)

There is little you can do in the short time between an announcement of an inspection and the actual inspection, which usually begins with a document review. However, in the time you do have, try to have a Health and Safety Program, Infection Control Plan for Tuberculosis, Exposure Control Plan for Bloodborne Pathogens, Ergonomic Policy, Workplace Violence Prevention Policy, Employee Evacuation Plan, Hazcom Program, and Lockout/Tagout. Having something, even a draft or copy of something, is better than nothing at all.

H.   LAST-MINUTE ROUNDS (SEE CHECKLIST)

Touch base with all staff and department heads when you know that an inspection is on the way, regardless of how much lag time there is (a few hours or a few days). Do a quick inspection for the obvious: machine guarding, container labeling, infectious waste container and handing, obstructed fire exits, extension cords, frayed wires and missing ground pins, electrical boxes, availability of personal protective equipment, lockout/tagout equipment and box, fire extinguisher placement and maintenance, eyewashes in working order, etc.

I.   LAST-MINUTE STAFF KNOWLEDGE BRUSH-UP (SEE CHECKLIST AND INTERVIEW GUIDE)

Test and brush up on staff knowledge and responses to questions about appropriate personal protective equipment use, bloodborne materials handling, hazard communication, Material Safety Data Sheets (MSDSs), lockout/tagout, ergonomics, lifting procedures, emergency evacuation, eyewashes, first aid treatment, fire extinguisher use, etc.

J.   PROVIDING A WORK SPACE FOR DOCUMENT INSPECTION AND EMPLOYEE INTERVIEWS

Except for rare occasions involving eminent danger, the inspector will ask to review the OSHA 200 Log first. For a regular inspection, the inspector will also want to review your written Health and Safety Program and all the mandatory policies and training records. Provide a quiet place where the inspector can conduct document reviews, and later, staff interviews.

K.   PROVIDING AMENITIES

Offer the inspector coffee or beverage and show the inspector where the rest rooms are and how to access them if keys are necessary. Offer the ordinary lunch. Do not be excessive. Inspectors are allowed to receive such normal offers from employers up to a certain point (approximately $20/day) before it can be questioned as bribery. On the other hand, do not make things so pleasant that the inspector may linger for the comforts. Many of these inspectors are accustomed to inspecting construction sites outdoors and other heavy manufacturing environments. It is a treat and a luxury for them to be in a heated or air-conditioned environment with a cushioned chair. So, provide basic amenities, but, again, no need to go over the top or offer the very best you have. Some inspectors joke that if they were provided a cold room and a hard chair to work in, the inspection would end more quickly.

L.   LIAISON-ACCOMPANIED INSPECTION

1.   Equipment

The liaison should be equipped with the following when accompanying the inspector on the site survey: clipboard or notepad and pens; camera; *keys* to gain entry into secured areas; and any personal protective equipment necessary (gloves, goggles, hard hats if walking through certain renovations or construction areas, etc.)

2. Scope of Facility Tour

The liaison should only take the inspector to inspect places that the inspector requests and that are within the scope of the intended inspection as set forth in the opening conference.

3. Answering Questions

The liaison should answer questions as limited as possible to the fundamental parameters of the inspection. In cases when the answer is unknown or unclear, rather than fortuitously creating a liability of "having knowledge" or "recognition of" and thus "negligence" or "willful," leading to the ability of the inspector to establish employer knowledge of a hazard and/or violation, which is then the basis of a citation, merely answer, "I don't know." Do not offer to ask or find someone else to answer the question. Let the inspector initiate all the inquiries.

4. Document Inspection

The liaison shall take notes and pictures during the inspection especially if there is any question about a certain piece of equipment, layout, arrangement, etc. At the closing conference review the inspector's notes along with your notes to ensure that you have everything that the inspector has. Remember, unlike DOH or joint commission surveys, OSHA inspectors are not to expand on their findings after their review during the closing conference.

M. Employee Interviews

The inspector can interview up to 13% of the employee population. The inspector will randomly select employees from different departments. (If there is a union, the shop steward may be asked to select the employees to be interviewed.) Interviews are to be conducted in private without management presence (see Chapter 5).

N. Following Up and Verifying Exchange of Information

Follow up with staff and department heads after they have had an interview with an OSHA inspector. Try to determine what was asked and how the staff responded. Your clarity on these interactions, although you are not permitted to be present, may help alleviate misconstrued citations that the inspector may create in the future, due to a misunderstanding or misinterpretation of your staff's intended responses.

O. Speaking Out Policy, or Making A Case

If upon inspection, the inspector notes something, which the inspector questions or finds to be a violation, but for which you have an alternative explanation, speak out at this point. Do not remain silent if you think you have a good standing argument why something does not pose a hazard, or does not pose as severe a hazard as the inspector believes. Respectfully make your point known. This is the one part of the inspection where providing additional information may be helpful to averting a citation, or a higher charge if one is issued.

Follow the basic tenants of OSHA citations listed below. In issuing a citation, an employer must have knowledge, but that knowledge is based on what is reasonably recognized, that a hazard exists, and is not abated. The degree of hazard speaks to the degree of the penalty and citation. Such arguments are more easily accomplished with issues that are not specifically addressed by a standard that would be cited under the General Duty Clause. There is no written rule denoting exactly what is expected if no standard exists. Therefore, it is easier in such cases to argue that no hazard actually exists, or your method of hazard abatement is satisfactory. However, even among issues that are regulated by unequivocal standards, arguments may still stand if alternative methods of hazard abatement provide *equal* or *superior protection.*

You have three main defenses during the inspection, and possibly afterward, if you receive a citation and choose to contest:

1. Hazard—Prove that a hazard does not in fact exist. Or, if the hazard does exist, prove that the degree of hazard is less than what the inspector deems it to be.
2. Abatement—Prove that the hazard has been abated satisfactorily.
3. Recognition—Prove that you could not have *reasonably known* about the hazard.

These are the technical bases of a citation, and any effective defense of one must address at least one of these.

P. WHEN TO COOPERATE AND WHEN TO REQUIRE SUBPOENAS

Cooperate to the extent of all the issues covered in this book. When OSHA requests fall outside of this designated area, and that of the following examples, then respectfully demur and have the agency subpoena those documents or activities. This will offer better protection from capricious requests and from any future liabilities should there be any problems, i.e., OSHA loses documents (which is not uncommon) or there is an incident, injuring a resident.

Q. PATIENT'S MEDICAL RECORDS

Inspectors do not usually ask to review patient's medical records. If they do, they are looking for tuberculosis status, lifting needs, etc. Since it is not customary for them to ask for such records, and your first priority is the privacy and well-being of the patients, it is probably best for you to ask the inspector to obtain releases from the families or individuals. There is no need for you to cooperate to this end, as it may conflict with your mandate. Explain this to the inspector courteously, and ask that the records be subpoened.

R. EMPLOYEE MEDICAL RECORDS

Inspectors have the right to review employee medical records. They may ask for copies of the records, which is also not customary. Again, have them subpoena this.

S. FISCAL RECORDS

Inspectors do not look at fiscal records, nor do they ask for a copy of them, in general. This issue has arisen out of questions about a facility's ability to afford lift equipment. However, this is beyond the usual standards and standard operating procedures of OSHA inspections. Again, have them subpoena this information if they require it.

T. VIDEOTAPING

Inspectors have used videotaping for ergonomic risk assessments. This may make more sense in a manufacturing setting. But in long-term care, it puts residents at risk since staff will be nervous. In addition, staff will probably not execute an incorrect lift before a camera. Explain the position that you must protect the well-being of the residents and that you cannot allow this to go forward, particularly when you see that little can be gained from it. Again, if they insist, have them issue a subpoena. Ensure that they obtain all the necessary releases from residents before they begin taping.

U. REMAINING CALM

The following applies particularly to unusually difficult inspectors who may seem antagonistic, biased, and manipulative. Do not let the process ruffle your or your staff's feathers. That is actually the highest price you will pay for their visit, if you let it happen. Calm your staff. Offer information and advice. Often, if OSHA cannot find anything substantive to cite, it may create the perception of serious issues and problems, when there is really little of consequence. The style of your interactions is dependent on the home office corporate culture, and the individual that was assigned to work with you. In these days of OSHA reform and Chamber of Commerce injunctions, there is a confused mandate and little uniformity on how they are to approach facilities. Some inspectors may take it upon themselves to continue the traditional style of intimidation, or they may take this opportunity to be more proactive and cooperative. In any case, you should remain conscientious and calm. Panic, fear, and hysteria are counterproductive to the process and your health.

On the other hand, if OSHA and employee safety and health are of little concern to your facility, then OSHA compliance is probably not your greatest liability, although you may be out of compliance. The commitment to and the quality of the safety and health system of any organization are reflective of the quality of the management system and ultimately affect overall performance. The greatest liability in ignoring these issues is the cost in terms of workers' compensation insurance and its indirect costs, staff turnover, burnout, and low morale. It would be a form of incomplete management. It is the author's goal to help facilities create and maintain a working health and safety management system that, by the way, also manages OSHA compliance issues within the context of a well-run system. This system shall be documented and evident in such a way that it would buy the facility "good faith" and result in good scores on the PEP so that the facility can also enjoy the benefits of a low-impact inspection, should one occur.

V. CLOSING CONFERENCE

At the end of the inspection, there should be a closing conference where the inspector should review the findings. It is not uncommon for facilities to go through an inspection and closing conference believing that all is well, and 2 weeks later be shocked by a citation in the mail. Make sure that you review the details of the inspector's notes and findings to ensure that you have the same. Address any issues that may still be in question as much as possible, including showing the inspector corrected hazards that were cited and abated during the inspection. You will still receive a citation, but it will reduce the paperwork for proving abatement. It will also later help to reduce the fine during the informal conference or hearings.

W. SEEKING HELP

If you wish to gain assistance at the start of an inspection, obviously, turn to your regular or available resources, in-house and regular safety and health consultants. If you turn to your insurance carrier's field representatives, you are better off seeking the help of your workers' compensation carrier, than, say, your fire/building/liability carrier. Be aware that most insurance field service representatives may be "loss control experts." However, they are not necessarily trained in OSHA compliance. Their training stems mostly from in-house carrier-centric issues of liability and writing lines of risks. Their expertise may also be more generic if the carrier covers several different lines of coverage, than if it only covers workers' compensation specifically. These representatives may be of some technical help in assessing hazards, degree of hazards, and proper abatement methods. But be wary of allowing them to handle your case entirely, as they may not be very familiar with the OSHA standards, or the agency's administrative process.

If you venture forth to find outside assistance, at this point it is best to seek a safety and health consulting company that specializes in OSHA compliance. You are also probably better off finding a local consultant who is familiar with and may have well-established relationships with the local

OSHA offices and their players. As in shopping for any and all types of consultants, you can ask to verify their qualifications and speak to past or current customers. You will probably negotiate a contract based on the extent of your citations and what would probably be needed to bring everything to a closure. Plan on paying approximately $1000 a day for work done.

It is usually unnecessary and more costly in most instances to involve attorneys at this point, or even after citations are received. It is best to dispose of or reduce citations and penalties based on technical issues, not legal issues, which are much more constrained. Attorneys should be consulted should you develop a federal case, a landmark-type case, unless your labor law firm is truly equipped and specializes in OSHA law. Such lawyers do exist, but they are not common.

III.   OSHA PROGRAM EVALUATION PROFILE

A.   INTRODUCTION

The PEP is a comprehensive benchmarking tool that accesses critical elements of a health and safety system. It is highly reflective of current trends, and the established principles of modern professional safety and health management. In this way, it is very different from any materials that have previously been issued by this agency. It does not address any particular OSHA standards at all. It is purely a management assessment tool.

There is a five-point scoring system. A 1 is a nonexistent safety program. A 2 is development. A 3 is basic. A 4 is superior. A 5 is outstanding. OSHA is looking for a 3, a basic rating. If the score is lower than 3, the inspector will embark on a traditional wall-to-wall inspection. If the score is 3 or better, the inspector will scale back the inspection and conduct only a "focused inspection" on relevant issues that did surface.

Upon closer review of the PEP, you will notice that it is divided into six sections, one of which contains only one subcategory, and one with four. It is uncertain how scoring will be weighted for each subcategory. It is something that OSHA is working on, since the values for certain subcategory scores will be diminish by mere fractionalization. These will most likely be weighted differently to equalize their values since, for example, "Employee Participation" and "Accident Investigation," "First Aid," and "Emergency Preparedness" are no less important than "Employee Training." (The body of the PEP, presented on page 32, is available on the CRC Web site as PEP.doc.)

B.   VALUE OF THE PEP TOOL

The author has found OSHA compliance efforts in the past to have relatively little impact on curbing workers' compensation losses. The author's mandate was the financial integrity of a workers' compensation insurance program, steady and growing dividends for participants, and a strong membership profile. This is achieved through selective membership, appropriate risk management, and effective health and safety resources and consulting to *improve overall safety performance.* Traditional OSHA enforcement methods were so inane as to have little effect on these comprehensive goals. The PEP, however, is a very timely tool in the author's opinion, to help the group, its members, and, now, any long-term care facility that uses it, along with the materials in this book, to achieve the goal of better overall performance resulting in lowered costs and increased stability. Nothing in this section is mandatory, per se. All of it is extremely helpful to implement the mandatory items, showing good faith, and running a good safety program.

C.   DESCRIPTION OF THE PEP

The following pages contain the entire body of the PEP. It has been slightly altered from its original form published by OSHA to render it more user-friendly. The front page is the comprehensive scoring sheet. It lists the six main headings, and all the subcategories therein, the score across the board for each subgroup, from 1 to 5, and the overall score. Each page after this initial cover page shows the scoring description for each subcategory. These definitions are reasonably clear, providing parameters for each score level in each subcategory.

The scoring guidelines are quite definitive. Upon reviewing them, it is credible that two different inspectors, given the same and accurate information, would result in strongly similar scores. However, readers will also notice that the only method in discerning scores properly for each component, or category and subcategory, is through a thorough review of critical elements of the safety management process.

D.   WRITTEN HEALTH AND SAFETY PROGRAM

A fundamental driver for all the components evaluated in the PEP begin with, and should be directed through, the written Health and Safety (H&S) Program. Having a written H&S Program in the past had afforded employers the old maximum 25% discount on their penalties for good faith. In the same vein, a comprehensive written H&S Program is the basis for much of the facets evaluated in the PEP. The payback in the pilot program is even greater. The PEP pilot program further legitimizes the concept that a good working safety system will grant the employer leniency and good faith discounts in that a basic score of 3 or better will determine that the inspector only conduct a focused inspection, whereas a failing score below 3 would lead to a traditional wall-to-wall inspection without leniency. Also, the good faith discount has been extended to 80% of penalties in the pilot program for adjustments. (There is a model sample H&S Policy for long-term care in Section IV.)

E.   EVALUATION AND SCORING

To conduct a proper evaluation and produce a valid score, the inspector must review documentation of safety management activities, primarily those that show evidence of the existence of a program of continuous improvement by self-assessment of risks, hazard identification, recommendation of corrective actions, follow-up of corrective actions, and follow-through to determine if those actions had the desired effect. This is reviewed along with the regular documentation of subject-specific policies, procedures, and training records (Ergonomics, Workplace Violence, Hazard Communication, Emergency Plan, Lockout/Tagout, Tuberculosis, etc). A walk-through and interviews with employees will be conducted to verify that the documentation is reflective of reality and to investigate employee concerns.

F.   HEALTH AND SAFETY ACTIVITIES LOG

A Health and Safety Activities Log is a chart that succinctly shows what issues have come up, when, what was discussed, what was done about it, when, by whom, and whether the issue is still pending or resolved. The Health and Safety Activities Log affords users the convenience of viewing progress and agenda *at a glance*. This is very much in keeping with the QA style of Joint Commission documents. It also affords you the ability to show inspectors, or anyone, your safety management activities without divulging names of residents or staff, as it is a highlighted summary.

To the end of presenting a credible safety management program, a facility needs to show *implementation* of the working safety system. The written Health and Safety Program, all the mandatory policies, and good in-service records (*see* Section IV), are only paperwork. The inspector needs to verify the safety program actually exists, and then score its effectiveness. Much of this is done through employee interviews and the site survey. The author suggests that you enhance this process by furnishing a "Showcase Log," a different version of the working H&S Activities Log containing only resolved issues, and activities about which you want to boast. Present it along with the written H&S Policy. This helps to alleviate the pressure on employee interviews to validate your theoretical documentation and policies (*see* Section V).

A facility would not *be well served by handing over their safety committee minutes to an OSHA inspector.* Safety committee minutes over time can establish the fact that an employer had recognized a hazard, had knowledge of its existence, but did not resolve the issue or resolve it in a timely manner, for whatever reasons. *Do not bring up the issue of safety committee minutes at all with the inspector.* OSHA inspectors generally will not ask for safety committee minutes. It is not ordinary protocol. However, if the inspector has been directed by his or her superiors to target your facility, and to go on a specific "witch hunt," to find something, especially a willful violation, then that would be about the only likely scenario where an inspector will ask to see safety committee minutes.

If the inspector asks to review your safety committee minutes, continue to be cooperative, friendly, and respectful. Explain that you would like to cooperate but that your safety committee minutes contain confidential patient information, which you unfortunately cannot share with the inspector. Continue, however, to divert the inspector's attention to the H&S Log, which is a summary of issues discussed at the meeting in a chart form. Explain that you would be glad to share this with the inspector, since it reflects the same basic issues and activities, but does not contain any patient information. Show the inspector the Showcase Health & Safety Activities Log, the one containing only items that have been resolved, and activities you would like to boast about.

Posting individual safety committee minutes in an employee area is encouraged as part of your communication process. This would not pose the type of liability that access to minutes over distinct time periods may. The benefits of posting individual safety committee minutes are far greater than the liability it may carry.

OSHA PROGRAM EVALUATION PROFILE

		Absent or Ineffective 1	Developmental 2	Basic 3	Superior 4	Outstanding 5	Score
I. Management Leadership and Employee Participation	A. Management Leadership						
	B. Employee Participation						
	C. Implementation						
	D. Contractor Safety						
II. Workplace Analysis	A. Survey and Hazard Analysis						
	B. Inspection						
	C. Reporting						
III. Accident and Record Analysis	A. Accident Investigation						
	B. Data Analysis						
IV. Hazard Prevention Control	A. Hazard Control						
	B. Maintenance						
	C. Medical Program						
V. Emergency Response	A. Emergency Preparedness						
	B. First Aid						
VI. Safety and Health Training	Training						
	Sum						
							Total Score

RATING REFERENCE

I. MANAGEMENT LEADERSHIP AND EMPLOYEE PARTICIPATION

A. MANAGEMENT LEADERSHIP

Visible management leadership provides the motivating force for an effective safety and health program.

1. Absent/Ineffective

- There are no health and safety policy, goals, objectives, or interest in safety and health issues at this worksite.

2. Developmental

- Management sets and communicates safety and health policy and goals, but is detached from all other safety and health efforts and activities.

3. Basic

- Management itself follows all safety and health rules, and gives visible support to the safety and health efforts of others.

4. Superior

- Management participates in significant aspects of the site's safety and health program, such as site inspections, incident reviews, and program reviews.
- Incentive programs that discourage reporting of accidents, symptom injuries, or hazards are absent.
- Other incentive programs that truly increase awareness, teamwork, good work practices, and proper reporting may be present.

5. Outstanding

- Site safety and health issues are regularly included on agendas of management operations meetings.
- Management clearly demonstrates—by involvement, support, and example—the primary importance of safety and health for everyone on the worksite.
- Performance is consistent and sustained or has improved over time.

B. EMPLOYEE PARTICIPATION

Employee participation provides the means through which workers identify hazards, recommend and monitor abatement, and otherwise participate in their own protection.

1. Absent/Ineffective

- Worker participation in workplace safety and health concerns is not encouraged.
- Incentive programs are present that have the effect of discouraging reporting of incidents injuries, potential hazards, or symptoms.
- Employee(s) representatives are not involved in the safety and health program.

2. Developmental

- Workers and their representatives can participate freely in safety and health activities at the worksite without fear of reprisal.
- Procedures are in place for communication between employer and workers on safety and health matters.
- Worker rights under the Occupational Safety and Health Act to refuse or stop work that they reasonably believe involves imminent danger are understood by workers and honored by management.
- Workers are paid while performing safety activities.

3. Basic

- Workers and their representatives are involved in the safety and health program, involved in inspection of work area, and are permitted to observe monitoring and receive results.
- Workers and representatives' right of access to information is understood by workers and recognized by management.
- A documented procedure is in place for raising complaints of hazards of discrimination and receiving timely employer responses.

4. Superior

- Workers and their representatives participate in workplace analysis, inspections and investigations, development of control strategies throughout facility, and have necessary training and education to participate in such activities.
- Workers and their representatives have access to all pertinent health and safety information, including safety reports and audits.
- Workers are informed of their right to refuse job assignments that pose serious hazards to themselves pending management response.

5. Outstanding

- Workers and their representatives participate fully in development of the safety and health programs and conduct of the training and education.
- Workers participate in audits, program reviews conducted by management or third parties, and collection of samples for monitoring purposes, and have necessary training and education to participate in such activities.
- Employer encourages and authorizes employees to stop activities that are potentially serious safety and health hazards.

C. IMPLEMENTATION

Means tools, provided by management, include:

- Budget
- Information
- Personnel
- Assigned responsibility

- Adequate expertise and authority
- Means to hold responsible persons accountable (line accountability)
- Program review procedures

1. Absent/Ineffective

- Tools to implement a safety and health program are inadequate or missing.

2. Developmental

- Some tools to implement a safety and health program are adequate and effectively used; others are ineffective or inadequate. Management assigns responsibility for implementing a site safety and health program to identified person(s).
- Management's designated representative has authority to direct abatement of hazards that can be corrected without major capital expenditure.

3. Basic

- Tools to implement a safety and health program are adequate, but are not all effectively used.
- Management representative has some expertise in hazard recognition and applicable OSHA requirements.
- Management keeps or has access to applicable OSHA standards at the facility, and seeks appropriate guidance information for interpretation of OSHA standards.
- Management representative has authority to order/purchase safety and health equipment.

4. Superior

- All tools to implement a safety and health program are more than adequate and effectively used.
- Written safety procedures, policies, and interpretations are updated based on reviews of the safety and health program.
- Safety and health expenditures, including training costs and personnel, are identified in the facility budget.
- Hazard abatement is an element in management performance evaluation.

5. Outstanding

- All tools necessary to implement a good safety and health program are more than adequate and effectively used.
- Management safety and health representative has expertise appropriate to facility size and process, and has access to professional advice when needed.
- Safety and health budgets and funding procedures are reviewed periodically for adequacy.

D. CONTRACTOR SAFETY

An effective safety and health program protects all personnel on the worksite, including the employees of contractors and subcontractors. It is the responsibility of management to address contractor safety.

1. Absent/Ineffective

- Management makes no provision to include contractors within the scope of the worksite safety and health program.

2. Developmental

- Management policy requires contractor to conform to OSHA regulations and other legal requirements.

3. Basic

- Management designates a representative to monitor contractor safety and health practices, and that individual has authority to stop contractor practices that expose host or contractor employees to hazards.
- Management informs contractor and employees of hazards present at the facility.

4. Superior

- Management investigates a contractor's safety and health record as one of the bidding criteria in selecting a vendor or contractor.

5. Outstanding

- The safety and health program of the site ensures protection of everyone employed at the worksite, i.e., regular full-time employees, contractors, temporary, and part-time employees.

II. WORKPLACE ANALYSIS

A. SURVEY AND HAZARD ANALYSIS

An effective, proactive safety and health program will seek to identify and analyze all hazards. In large or complex workplaces, components of such analysis are the comprehensive survey and analyses of job hazards and changes in conditions.

1. Absent/Ineffective

- No system or requirement exists for hazard review of planned/changed/new operations.
- There is no evidence of a comprehensive survey for safety or health hazards or for routine job hazard analysis.

2. **Developmental**

- Surveys for violations of standards are conducted by knowledgeable person(s), but only in response to accidents or complaints.
- The employer has identified principal OSHA standards that apply to the worksite.

3. **Basic**

- Process, task, and environmental surveys are conducted by knowledgeable person(s), but only in response to accidents or complaints.
- The employer has identified principal OSHA standards that apply to the worksite.

4. **Superior**

- Methodical surveys are conducted periodically and drive appropriate corrective action.
- Initial surveys are conducted by a qualified professional.
- Current hazard analyses are documented for all work areas and are communicated and available to all the workforce; knowledgeable persons review all planned/changed/new facilities, processes, materials, or equipment.

5. **Outstanding**

- Regular surveys including documented comprehensive workplace hazard evaluations are conducted by a certified safety and health professional or a professional engineer, etc.
- Corrective action is documented and hazard inventories are updated. Hazard analysis is integrated into the design, development, implementation, and changing of all processes and work practices.

B. **INSPECTION**

To identify new or previously missed hazards and failures in hazard controls, an effective safety and health program will include regular site inspections.

1. **Absent/Ineffective**

- No routine physical inspection of the workplace and equipment is conducted.

2. **Developmental**

- Supervisors dedicate time to observing work practices and other safety and health conditions in work areas where they have responsibility.

3. **Basic**

- Competent personnel conduct inspections with appropriate involvement of employees.
- Items in need of correction are documented.
- Inspections include compliance with relevant OSHA standards.
- Time periods for correction are set.

4. **Superior**

- Inspections are conducted by specifically trained employees, and all items are corrected promptly and appropriately.
- Workplace inspections are planned, with key observations or check points defined and results documented.
- Persons conducting inspections have specific training in hazard identification applicable to the facility.
- Corrections are documented through follow-up inspections.
- Results are available to workers.

5. **Outstanding**

- Inspections are planned and overseen by certified safety or health professionals.
- Statistically valid random audits of compliance with all elements of the safety and health program are conducted.
- Observations are analyzed to evaluate progress.

C. **HAZARD REPORTING**

A reliable hazard reporting system enables employees, without fear of reprisal, to notify management of conditions that appear hazardous and to receive timely and appropriate responses.

1. **Absent/Ineffective**

- No formal hazard reporting system exists, or employees are reluctant to report hazards.

2. **Developmental**

- Employees are instructed to report hazards to management.
- Supervisors are instructed and are aware of a procedure for evaluating and responding to such reports.
- Employees use the system with no risk of reprisals.

3. **Basic**

    ▪ A formal system for hazard reporting exists.
    ▪ Employee reports of hazard are documented, corrective action is scheduled, and records maintained.

4. **Superior**

    ▪ Employees are periodically instructed in hazard identification and reporting procedures.
    ▪ Management conducts surveys of employee observations of hazards to ensure that the system is working.
    ▪ Results are documented.

5. **Outstanding**

    ▪ Management responds to reports of hazards in writing within specified time frames.
    ▪ The workforce readily identifies and self-corrects hazards; they are supported by management when they do so.

III.   ACCIDENT INVESTIGATION

A.   ACCIDENT INVESTIGATION

An effective program will provide for investigation of accidents and "near miss" incidents, so that their causes and means for their prevention are identified.

1. **Absent/Ineffective**

    ▪ No investigation of accidents, injuries, near misses, or other incidents are conducted.

2. **Developmental**

    ▪ Some investigation of accidents takes place, but root cause may not be identified, and correction may be inconsistent.
    ▪ Supervisors prepare injury reports for lost time cases.

3. **Basic**

    ▪ OSHA-101 is completed for all recordable incidents.

- Reports are generally prepared with cause identification and corrective measures prescribed.

4. Superior

- OSHA-recordable incidents are always investigated, and effective prevention is implemented.
- Reports and recommendations are available to employees.
- Quality and completeness of investigations are systematically reviewed by trained safety personnel.

5. Outstanding

- All loss-producing accidents and "near-misses" are investigated for root causes by teams or individuals that include trained safety personnel and employees.

B. DATA ANALYSIS

An effective program will analyze injury and illness records for indications of sources and locations of hazards, and jobs that experience higher numbers of injuries. By analyzing injury and illness trends over time, patterns with common causes can be identified and prevented.

1. Absent/Ineffective

- Little or no analysis of injury/illness records; records (OSHA 200/101, exposure monitoring) are not kept or conducted.

2. Developmental

- Data is collected and analyzed, but not widely used for prevention. OSHA-101 is completed for all recordable cases.
- Exposure records and analyses are organized and are available to safety personnel.

3. Basic

- Injury/illness logs and exposure records are kept correctly, are audited by facility personnel, and are essentially accurate and compete.
- Rates are calculated to identify high-risk areas and jobs.
- Workers' compensation claim records are analyzed and the results used in the program.
- Significant analytical findings are used for prevention.

4. Superior

- Employer can identify the frequent and most severe problem areas, the high-risk areas and job classifications, and any exposures responsible for OSHA-recordable cases.
- Data are fully analyzed and effectively communicated to employees.
- Illness/injury are audited and certified by a responsible person.

5. Outstanding

- All levels of management and the workforce are aware of results of data analyses and resulting preventive activity.
- External audits of accuracy of injury and illness data, including review of all available data sources are conducted.
- Scientific analysis of health information, including nonoccupational databases, is included where appropriate in the program.

IV. HAZARD PREVENTION CONTROL

A. HAZARD CONTROL

Workforce exposure to all current and potential hazards should be prevented or controlled by using engineering controls wherever feasible and appropriate work practices and administrative controls, and personal protective equipment (PPE).

1. Absent/Ineffective

- Hazard control is seriously lacking or absent from the facility.

2. Developmental

- Hazard controls are generally in place, but effectiveness and completeness vary.
- Serious hazards may still exist.
- Employer has achieved general compliance with applicable OSHA standards regarding hazards with a significant probability of causing serious physical harm.
- Hazards that have caused past injuries in the facility have been corrected.

3. Basic

- Appropriate controls (engineering, work practice, and administrative controls, and PPE) are in place for significant hazards.
- Some serious hazards may exist.
- Employer is generally in compliance with voluntary standards, industry practices, and manufacturers' and suppliers' safety recommendations.
- Documented reviews of needs for machine guarding, energy lockout, ergonomics, materials handling, bloodborne pathogens, confined space, hazard communication, and other generally applicable standards have been conducted.
- The overall program tolerates occasional deviations.

4. Superior

- Hazard controls are fully in place and are known and supported by the workforce.
- Few serious hazards exist.
- The employer requires strict and complete compliance with all OSHA, consensus, and industry standards and recommendations.
- All deviations are identified and causes determined.

5. Outstanding

- Hazard controls are fully in place and continually improved upon based on workplace experience and general knowledge.
- Documented reviews of needs are conducted by certified health and safety professionals or professional engineers, etc.

B. MAINTENANCE

An effective safety and health program will provide for facility and equipment maintenance, so that hazard breakdowns are prevented.

1. Absent/Ineffective

- No preventive maintenance program is in place; breakdown maintenance is the rule.

2. Developmental

- There is a preventive maintenance schedule, but it does not cover everything and may be allowed to slide or performance is not documented.
- Safety devices on machinery and equipment are generally checked before each production shift.

3. Basic

- A preventive maintenance schedule is implemented for areas where it is most needed; it is followed under normal circumstances.
- Manufacturers' and industry recommendations and consensus standards for maintenance frequency are compiled with.
- Breakdown repairs for safety-related items are expedited.
- Safety device checks are documented.
- Ventilation system function is observed periodically.

4. Superior

- The employer has effectively implemented a preventive maintenance schedule that applies to all equipment.
- Facility experience is used to improve safety-related preventive maintenance scheduling.

5. Outstanding

- There is a comprehensive safety and preventive maintenance program that maximizes equipment reliability.

C. MEDICAL PROGRAM

An effective safety and health program will include a suitable medical program where it is appropriate for the size and nature of the workplace and its hazards.

1. Absent/Ineffective

- Employer is unaware of, or unresponsive to, medical needs. Required medical surveillance, monitoring, and reporting are absent or inadequate.

2. Developmental

- Required medical surveillance, monitoring, removal, and reporting responsibilities for applicable standards are assigned and carried out, but results may be incomplete or inadequate.

3. Basic

- Medical surveillance, removal, monitoring, and reporting comply with applicable standards.

- Employees report early signs/symptoms of job-related injury or illness and receive appropriate treatment.

4. Superior

- Health care providers provide follow-up on employee treatment protocols and are involved in hazard identification and control in the workplace.
- Medical surveillance addresses conditions not covered by specific standards.
- Employee concerns about medical treatment are documented and responded to.

5. Outstanding

- Health care providers are on site for all production shifts and are involved in hazard identification and training.
- Health care providers periodically observe the work areas and activities and are fully involved in hazard identification and training.

V. EMERGENCY RESPONSE

A. EMERGENCY PREPAREDNESS

There should be appropriate planning, training/drills, and equipment for response to emergencies. *Note:* In some facilities the employer plan is to evacuate and call the fire department. In such cases, only applicable items listed below should be considered.

1. Absent/Ineffective

- Little or no effective effort to prepare for emergencies.

2. Developmental

- Emergency response plans for fire, chemical, and weather emergencies as required by 29 CFR 1910.38, 1910.120, or 1926.35 are present.
- Training is conducted as required by the applicable standard.
- Some deficiencies may exist.

3. Basic

- Emergency response plans have been prepared by persons with specific training.
- Appropriate alarm systems are present.
- Employees are trained in emergency procedures.
- The emergency response extends to spills and incidents in routine production.
- Adequate supply of spill control and PPE appropriate to hazards on site is available.

4. Superior

- Evacuation drills are conducted no less often than annually.
- The plan is reviewed by a qualified safety and health professional.

5. Outstanding

- Designated emergency response team with adequate training is on site.
- All potential emergencies have been identified.
- Plan is reviewed by the local fire department.
- Plan and performance are reevaluated at least annually and after each significant incident.
- Procedures for terminating an emergency response condition are clearly defined.

B.   FIRST AID

First aid/emergency care should be readily available to minimize harm if an injury or illness occurs.

1.   Absent/Ineffective

- Neither on-site or nearby community aid (e.g., emergency room) can be ensured.

2.   Developmental

- Either on-site or nearby community aid is available on every shift.

3.   Basic

- Personnel with appropriate first aid skills commensurate with likely hazards in the work-place and as required by OSHA standards (e.g., 1910.151, 1926.23) are available.
- Management documents and evaluates response time on a continuing basis.

4.   Superior

- Personnel with *certified* first aid skills are always available on site; their level of training is appropriate to the hazards of the work being done.
- Adequacy of first aid is formally reviewed after significant incidents.

5.   Outstanding

- Personnel trained in advanced first aid and/or emergency medical care are always available on site.
- In larger facilities a health care provider is on site for each production shift.

VI.   SAFETY AND HEALTH TRAINING

Safety and health training should cover the safety and health responsibilities of all personnel who work at the site or affect its operations. It is most effective when incorporated into other training about performance requirements and job practices. It should include all subjects and areas necessary to address the hazards at the site.

1.   Absent/Ineffective

- Facility depends on experience and peer training to meet needs.
- Managers/supervisors demonstrate little or no involvement in safety and health training responsibilities.

2.   Development

- Some orientation training is given to new hires.
- Some safety training materials (e.g., pamphlets, posters, videotapes) are available or are used periodically at safety meetings, but there is little or no documentation of training or assessment of worker knowledge in this area. Managers generally demonstrate awareness of safety and health responsibilities, but have limited training themselves or involvement in the training program at the site.

3.   Basic

- Training includes OSHA rights and access to information.
- Training required by applicable standards is provided to all site employees.

- Supervisors and managers attend training in all subjects provided to employees under their direction.
- Employees can generally demonstrate the skills/knowledge necessary to perform their jobs safely.
- Records of training are kept and training is evaluated to ensure that it is effective.

4. Superior

- Knowledgeable persons conduct safety and health training that is scheduled, assessed, and documented, and that addresses all necessary technical topics.
- Employees are trained to recognize hazards, violations of OSHA standards, and facility practices.
- Employees are trained to report violations to management.
- All site employees—including supervisors and managers—can generally demonstrate preparedness for participation in the overall safety and health program.
- There are easily retrievable scheduling and record-keeping systems.

5. Outstanding

- Knowledgeable persons conduct safety and health training that is scheduled, assessed, and documented.
- Training covers all necessary topics and situations, and includes all persons working at the site (hourly employees, supervisors, managers, contractors, part-time and temporary employees).
- Employees are trained to recognize inadequate responses to reported program violations.
- Retrievable record-keeping system provides for appropriate retraining, makeup training, and modifications to training as the result of evaluations.

IV. SAMPLE POLICY: HEALTH AND SAFETY POLICY FOR LONG-TERM CARE FACILITIES*

A. MISSION STATEMENT

The Management of _____ Nursing Home is committed to providing excellence in long-term care. Health and safety is an integral part of overall operational performance. The management strives to provide effective leadership in maintaining a safe and healthful workplace and working conditions for our staff, as it does in maintaining a safe and healthful home and living conditions for our residents. Health, safety, and quality are organization priorities.

B. PHILOSOPHY

The management recognizes safety as a quality issue. It realizes that meaningful resources devoted to health and safety are not a loss factor, but are actually investments that yield higher productivity and efficiency while curbing losses and limiting liability exposures. Management also realizes that it does, actively or passively, consciously or unconsciously, set the level of tolerable and intolerable safe and unsafe work practices and activities within the facility, in the same way that it sets the level of expected standards of performance and care, and manages accordingly.

---

*See, "Health&SafetyPolicy.doc," available on the CRC Web site.

C. LEADERSHIP

It is a management goal to produce and maintain an effective Health and Safety (H&S) Program to promote continually the highest level of health, safety, and operational performance possible. One of its results should be to limit, to the lowest degree possible, risks of accidents, injuries, near misses, and chronic or acute health exposures. The management has set an action plan to meet these goals, as delineated within this policy.

D. COMMUNICATION

Management communicates the goals of this H&S Program to all members of the organization in order that they understand the desired results and the action plan for achieving them. This is accomplished through chain-of-command communications (messages relayed and enforced through management and supervisors); open channel communications (upstream and downstream); new employee orientation; periodic training; employee participation in Safety Committee and other H&S Program activities; newsletter; memoranda; etc. The management is current on, and is greatly interested in, the activities and issues of the H&S Program.

The management strives to understand the nature of the safety culture in the facility ("How things are done here"), and the reasons for it. This is facilitated by periodic employee perception surveys, upstream performance surveys, an active suggestion box, and encouraging an open-door policy at all levels of management.

E. RESPONSIBILITIES, ACCOUNTABILITY, AND COMMITMENT

Management assigns responsibilities for developing and managing a comprehensive program to a person with expertise in occupational health and safety management, the Safety or Risk Manager. The Safety Manager is _____ of the _____ Department. The management grants the Safety/Risk Manager decision-making powers through designated procedures set forth in this policy.

Management goes on to endorse and support this H&S Program with financial, human, and material resources needed to achieve its goal. Management also provides for and mandates manager, supervisor, and employee accountability and responsibility for performing their assigned roles in the program.

The management establishes, communicates, and enforces a reward/commendation and disciplinary system that applies equally to all employees (managers, supervisors, and staff) who promote, enhance, maintain, break, or disregard safety rules, safe work practices, and procedures. Employees are encouraged to nominate staff, supervisors, and managers for commendations. Observations and complaints about poor work practices will be investigated and corrected immediately.

The chief executive of the facility is the chair of the Safety Committee. Or, in instances where employee involvement is very strong and well structured, the chair of such a committee can be a front-line employee, who reports directly to the chief executive of the facility. Whichever is the case, the chief executive of the facility is very closely tied to the activities of this committee. This committee serves the needs of the organization by facilitating employee involvement, feedback, communication, and organizational coordination and consensus.

F. EMPLOYEE INVOLVEMENT

Management selects staff for assignments in the H&S Program and on the H&S committee based on their special interests and/or expertise. Members may also be selected based on a pool of volunteers or departmental nominations. Tenureship on committees should be long enough to allow the members to become viable participants, able to provide meaningful interaction. They are routinely reviewed and shifted in response to signs of burnout and to give opportunities to other interested parties.

It is the policy of this facility that management provide employees who have expressed health and safety suggestions or concerns, both formal and informal, with a timely and reasonable response and follow-up. Management recognizes, commends, and awards employees, supervisors, and management staff who have made significant, continuous, or even subtle contributions to the health and safety effort. Employee feedback and involvement is strongly encouraged.

G.   WRITTEN HEALTH AND SAFETY PROGRAM

This written H&S Program has been tailored to the facility's mission and goals. This written program establishes clear objectives and an action plan, communicates health and safety policies, procedures, and protocols. It assigns responsibilities for the implementation of the program. This written program is reviewed regularly and updated and revised as needed.

H.   HEALTH AND SAFETY COMMITTEE

The Health and Safety Committee develops the mission, function, and specific goals for the facility H&S Program. Its function is to create and promote facility-wide awareness and concern for safety. It is the facilitator of safety performance for the facility. It should serve as a safety resource and advisory group.

The committee facilitates the implementation of the program elements. It reviews and analyzes all forms of communication relating to health and safety performance and promotion, including suggestions, complaints, and reports related to the H&S program activities. It makes recommendations related to program development, implementation, and revision. The committee may act as a liaison to help departments coordinate services and activities to suit better each unit's needs. The committee monitors the effectiveness of the program and reports its activities to the chief executive on a regular basis.

The committee has the authority to request information necessary to meet its mandate from all departments. It has the authority and responsibility to follow up on work orders; injuries and claims; work practices; enforcement or enhancement of existing rules and procedures; incident investigations; and requested revisions and changes, both procedural and physical.

The committee members include management and employees of all departments within the facility. The committee reports directly to the chief executive officer, or the senior administrator of the facility. It communicates its rationale and plans to the employees of the facility, for its actions and decisions not to act on particular issues. The committee expedites its duties in a timely and reasonable fashion.

The committee meets at a frequency of _____ times per _____. This meeting interval has been determined to be adequate but not fixed. Meetings can be called at more frequent intervals if the group is intensely working on something with time constraints or has some unusually pressing matters at hand.

I.   INCIDENT REPORTING

It is the policy that employees must promptly and accurately report inefficiencies, hazards, unsafe work practices, occupational injuries and illnesses, health and safety problems, and suggestions. In addition, they must be able to do so without fear of reprisal. This is expected procedure. Supervisors and management staff are trained to handle, properly complete and submit, encourage, and promote such reports. This information will be reviewed by the H&S committee to understand better the nature of hazards and problem procedures, to correct them, and to plan to prevent injuries.

J.   INCENTIVE PROGRAMS

[Include the body of the facility incentive program here, if one exists. Or make reference to it, stating its goals, and basic components. Input from the H&S committee should be part of the incentive programs. Incentive programs should be based on desired behaviors, not injury statistics.]

K. PERFORMANCE EVALUATIONS

[Make reference here to how safety performance, improved, exceptional, or poor, will impact on an individual's performance evaluations. Performance evaluations should be both formal and informal. They are marked by meaningful annual evaluations, which should be based on the informal daily routine of supervisory observation and feedback on work practices.]

L. POSTINGS REQUIREMENTS

OSHA, DOL, and other workplace posters are displayed in a prominent location where all employees are likely to see them. Emergency telephone numbers are posted in appropriate locations where they can be readily found in the event of an emergency.

Appropriate information concerning employee access to medical and exposure records and Material Safety Data Sheets (MSDSs), etc. are posted and/or otherwise made readily available to affected employees.

Signs concerning "exiting from buildings," room capacities, floor loading, exposure to microwave, noise, heat, chemical storage, medical waste storage, isolation, or other harmful substances are posted where applicable.

The summary of the previous year's OSHA 200 Log (the bottom left-hand corner of the last page of the OSHA 200 Log) is posted for the month of February.

M. RECORD KEEPING

1. H&S Committee Minutes

The minutes of the H&S committee are kept. They are posted and circulated throughout the facility in order that employees have the opportunity to be informed about its activities. Minutes are also shared with visiting consultants and made available for OSHA, survey, or Joint Commission review. These records are kept by ——————— [title] at the ——————— [office].

2. OSHA Log

All occupational illnesses and injuries, except minor injuries requiring only first aid (as defined by OSHA), are recorded as required on the OSHA 200 Log. Needle sticks are recordable on the OSHA 200 Log. Clean needle sticks will be regarded as a puncture wound and should be recorded accordingly if necessary, but shall not be regarded as a potential exposure. A tuberculosis conversion from a previous negative to a positive on an annual baseline test is recordable on the OSHA 200 Log. (This does not apply to new hires.) The Log is kept by ——————— [title] of the ——————— [office].

3. Incident Report Forms

There are separate and distinct reporting forms for employees and residents/visitors because they may develop into workers' compensation or liability claims and may involve OSHA or DOH scrutiny, respectively. Thus, they are distinctly different in nature and require different approaches and documentation.

The accident and incident records are kept in an well-organized fashion to facilitate information and status updates, as well as future investigations or data collection. The employee incident report forms contain appropriate questions to comply with the OSHA 101 report form, gain important accident investigation insight, and collect necessary workers' compensation information. The resident/visitor report forms should be adequate to meet the needs and requirements for forwarding and managing liability claims, and DOH reviews and investigations.

Accident/incident records are complete, with meaningful information to convey important details of what actually happened or is alleged to have happened. Supervisors and managers are trained on the importance of and the proper completion of such reports.

Accident reports include a section where the employee signs an acknowledgment clause that any intent to falsify any information on this report is against the law and is punishable by law.

The completed form is routed to _____ [title] of the _____ [office], who compiles a monthly/quarterly report for H&S committee review. These results are to be viewed as indicators of trends. They shall not be used as the sole measure of a department or the facility's safety performance. Incident records are only reflective of indirect downstream factors. Incidents may continue to exist and reporting may even be increased during positive changes in a safety culture. Meanwhile, incidents may be absent due to probability and/or underreporting when little or no attention has been given to the effort, and safety performance and the culture may be deteriorating. Thus, incident records should be carefully interpreted and used only as one of several tools to benchmark safety performance.

4. Workers' Compensation Records

There are clear records of certificates of workers' compensation insurance for all contractors and vendors who enter the grounds of the facility and provide services. This includes contract physicians and other health care professionals. In such cases, the facility will verify that the coverage does not exempt the owner/operator/president or vice president of the company, which may in fact be the professional who provides services to the facility.

5. Employee Access to Medical and Exposure Records

This facility notifies employees of their right to access medical or exposure records. This is accomplished by [posted notices, or hand-delivered notices via paycheck stuffers, or orientation, etc.] A copy of the OSHA standard requiring this is available at the _____ [office] for employee and inspector review.

The facility will furnish the employee with the requested information within 15 days, at no cost to the employee, at a reasonable time and place. Records include accident/incident reports; MSDSs, any reports of collected data of which the employee was part of the data set. Access will also be granted to representatives of the employee with written permission from the employee.

6. Record Retention

All employee medical records and records of employee accidents, and exposures to hazardous substances or harmful physical agents including MSDSs are kept up-to-date. Arrangements have been made to maintain required records for the legal period of time for each specific type of record (30 years after employee's term with the facility).

7. Training Records

Training records are properly kept and include the name and qualification of the instructors, content of in-services, training time, date, and complete roster with sign-in sheet, handouts, course objectives, and the results of any quizzes or tests.

8. Building and Equipment Records

Operating permits and records are kept up-to-date for such items as elevators, air pressure tanks, gas tanks, air stacks, well water tests, etc.

Clear and complete records of manifests for medical waste and hazardous waste shipments are kept. There are clear records establishing the legality, technical ability, and the insurance status and permit to operate status, of treatment facilities that the facility sends its medical and other hazardous wastes to. This is the responsibility of the director of maintenance.

9. Mandatory Policies

All required programs are written, updated, and reviewed to include all required components: Hazard Communication; Electrical Work Practices/Lockout/Tagout; Bloodborne Pathogens/Universal Precautions; Tuberculosis/Infection Control; Emergency Plan; Personal Protective Equipment; Ergonomic; Workplace Violence Prevention; etc. (For further assistance in these policies, please contact our office or refer to our *OSHA Compliance for Long-Term Care Manual.*)

10. Additional Policies

[Include or make references to other related policies and procedures. Make sure these policies include safety performance in their description and assessments to *ensure that safety is an integral part of overall operations* and is accounted for in the following job descriptions, training needs assessments; general work practices; job- and task-specific work practices; department specific policies and procedures, etc. For additional guidance, please refer to Chapters 3 and 4.]

Reviewed by:

Name	Title	___/___/___ date
Name	Title	___/___/___ date
Name	Title	___/___/___ date
Name	Title	___/___/___ date

V. OSHA PREPAREDNESS: HEALTH AND SAFETY ACTIVITIES LOG

The following is a simplified record-keeping device designed to organize health and safety concerns and activities in a user- and inspection-friendly fashion. This guide is further designed to assist your facility in garnering *as much credit as possible* for activities already taking place, which very often inadvertently evade the health and safety activities document trail.*

A. SAFETY COMMITTEE USERS

Figure 2.1 is a good device for safety committee to keep track of its findings and activities. It serves as an excellent supplement to committee minutes to present the issues, discussions, actions, and status of any given subject, at a glance.

B. SAFETY PROGRAM DOCUMENTATION

Often, activities that take place throughout a facility impact on safety performance, and are in fact a part of safety management. These "operational" elements, where safety management is actually occurring as part of the system, may not necessarily be expressed through the safety committee or safety-specific programs. Without a keen screening system to document and credit yourself with safety management activities, you may be hard pressed to show full evidence of your safety stewardship, even though there are operational benefits and safety performance will ultimately increase. The following are some common examples of safety management activities that often go undocumented under safety activities or programs, but reflect excellent safety stewardship, nonetheless.

---

*Available on the CRC Web site, *"Health&SafetyActivitiesLog.doc,"* and *"Health&SafetyActivitiesLog.xls."*

Item	Date	Subject	Discussion	Conclusion	Action	By	Follow-Up	Pending	Resolved
#									
#									
#									
#									
#									
#									

Figure 2.1   One should keep this log of items that are resolved, or expected to be resolved shortly, to show the inspector. A complete log of all issues should be kept with the safety committee minutes that would not be shared with an inspector.

Example 1

Nursing decides to go with a self-sheathing needle system or a needleless system to cut down on the number of needle sticks or lancet cuts. Upon its introduction, nursing realizes a considerable return in savings from decreased cost of time lost and medical costs avoided, although the inventory may have cost upward of sevenfold at first. There was a process of hazard evaluation/ assessment, abatement, and follow-up. However, if this were treated as a "nursing" issue, that may or may not have even been discussed or documented through QA. This important safety management activity may not even be noted when superficial safety inspections are conducted by less-concerned or less-knowledgeable agents if it is not also recorded within the document inspection path.

Example 2

The director of housekeeping and/or dietary reviews the list of chemical inventory. The director applies chemical safety management principles and decides that a different combination of cleaners would do the same job, but with less toxic materials, or less quantity of stored materials on site, or a fewer number of different types of materials, or less to no probability of interreactivity of chemicals present. This notable process of safety management may not be recorded anywhere, never mind via the official safety system.

Example 3

Residents' behaviors are monitored and discussed. A usually docile and cooperative resident may be noted as becoming more irritable and unpredictable. Care plans are reviewed and adjusted to reflect this change for caregivers to be more aware and prepared to deal with the resident. Medical review is conducted to consider potential physical and psychological causes. A process of trial and error is engaged in that may include a variety of medical treatments, medications, behavior-modification programming, etc. to help alleviate the behavior(s). The focus of these activities is resident well-being, of course, that, in turn, would increase the well-being of the residents around this individual and the staff. This is, in essence, a process of safety management, to prevent "workplace violence" in that it works to decrease the probability of an employee being struck and injured by a combative or assaultive resident. The hazard is, in fact, evaluated/monitored, abatement methods are developed and implemented, and follow-up is conducted to determine efficacy. How often are such activities properly recorded in safety programs and activities documents? These are often recorded in care plans and QA meetings, but, perhaps, nowhere else that a safety inspection agent may have access to, or even think to ask about.

Example 4

The same process occurs as in the preceding example, except that the issue is not behavior, but the ability of a resident to be weight-bearing or not, or less so, has been reported to wane, and the change is noted, monitored, and accounted for. Again, this process results in many benefits to the well-being of the resident. But, also, if done consistently and aggressively, this is part of a strong safety management system that prevents employee injuries during patient transfers.

Example 5

A maintenance director secures the door to the elevator shafts and places an approved permanent sign on the door, "Confined Space—Restricted Entry."

Example 6

A maintenance supervisor notices frayed wires and missing grounds on plugs periodically on in-house electrical appliances. He suggests a low-impact system of color tagging all cords and appliances throughout a given year to ensure that each piece of equipment is inspected at least annually. His idea is implemented and becomes part of the preventative maintenance plan and activities of his department.

Example 7

Team leaders give each shift a short summary to each shift before they begin to brief them on the status of the residents and activities on the floor. The report includes any changes in care plans (lifting requirements), comments from the last shift about the volatility or moods of certain residents, the operational status and whereabouts of lift equipment or personal protective equipment, or other devices that improve safety performance, workflow, etc. Sharing information from remarks in a shift log is comparable to this.

Example 8

An orientation on workers' compensation insurance, costs, and a review of the claims is provided to management staff or regular staff to provide them with an understanding of the problems in the facility, and its personal and financial impact on the injured employees, remaining staff, and the institution.

C. ACTUAL SAFETY PROCESS VS. INSPECTION ELEMENTS OF THE SAFETY PROCESS

The irony is that, if safety is truly managed in the optimal way, that is, as integrally part of the system of operations, there would be little evidence recorded through a system devoted entirely to safety, since the information would be absorbed and materially part of all other activities.

D. OPTIMAL USE OF THE HEALTH AND SAFETY ACTIVITIES LOG

This Health and Safety Activities Log and this guide are designed for in-house facility use to encompass as much of those activities already ingrained through operations that are also safety management activities. The basic tenants of safety management are:

Monitoring/evaluation/assessment of hazards
Development and implementation of a solution or "abatement method"
Follow-up actions to ensure resolution of the issue and/or to verify the effectiveness of the abatement method

If you find any management staff or regular staff engaged in any of the above, document it on the Log. The following is a more detailed list of common activities that are part of this safety management process:

Communicating Downstream—Not only about policies and procedures, but resident care (infection control, workplace violence prevention, etc.)

Communicating Upstream—Employee feedback

Providing Feedback—Work procedures, performance, including positive

Training—Both formal, and informal one-on-one feedback and demonstration

QA—In assessing staff knowledge of protocols, and consistency of practice

Monitoring, benchmarking, evaluating or assessing performance

E.   WHAT SUBJECTS TO REPORT ON

Subjects to be reported on, of course, should not be limited to any given laundry list so long as it falls within the realm of safety management as discussed above. Some common subjects to report on include but are not limited to the following. This list may assist you and your management staff to begin to identify those typical operational activities that are automatically engaged, that should be reported on this Log in order that you garner as much credit as possible for safety management activities:

Machine guarding

Slips, trips, and falls

Ergonomics—body mechanics/proper lifting design, and techniques

Lockout/Tagout—electrical energy isolation

Electrical safety

Fire safety

Hazard communication—chemical safety

Bloodborne pathogens—infection control

Tuberculosis—infection control

Workplace violence prevention—safety and security

Workplace violence prevention—resident behavior management

Confined spaces (securing potentially dangerous areas from regular or employee entry, entry by authorized personnel only)

OSHA 200 Log—review of employee work-related injuries and illnesses

Employee participation in safety—formal systems of feedback, meaningful participation (with appropriate training) in safety teams, or committees

F.   WHICH WORDS TO USE

When using this Health and Safety Activities Log, or recording health and safety activities (i.e., training), it would be advisable to learn to begin to use OSHA terminology, along with your own industry and in-house verbiage to describe the subjects. The categories above are the OSHA correct titles for the various subjects. This will definitely help expedite an inspection, and help you and your staff to be more versed in OSHA terminology when holding discussions with the inspector, and when being interviewed by the inspector.

G.   WHAT ACTIVITIES TO LOG

When attempting to determine whether an activity within a subject should be reported on this log, consider if the activity decreases exposure to the potential hazard, or decreases probability of injury or illness, and/or if the activity increases health, safety, awareness, or better work practices regarding the same subject. For example:

Ergonomics—changes or activities that decrease repetitive motions, exposure to poor posture, or encourage poor body mechanics or poor posture

Ergonomics—Changes or activities that increase better body mechanics (i.e., proper materials storage and design of storage areas, use of materials handling devices, etc.)

Hazard Communication—Changes or activities that decrease chemical inventory, chemical storage amounts, number of chemicals present, reactivity of substances

Hazard Communication—Changes or activities that increase chemical hygiene practices (providing appropriate personal protective equipment, training, enforcement including positive reinforcement, constructive feedback, counseling, disciplining)

Workplace Violence Prevention—Changes or activities that decrease probability of an employee sustaining injuries from combative residents

Bloodborne Pathogens/Tuberculosis–Infection Control—Changes or activities that decrease the probability of employee exposure to bloodborne pathogens or tuberculosis and/or increase and reinforce proper work procedures and awareness

Employee Participation—Any activities that encourage employee participation in learning, inspecting, monitoring, reviewing losses and findings, identifying hazards, determining solutions, abatement, following up, etc.

H. OTHER PRACTICAL NOTES ON USING THE LOG

A safety committee could be the driver for all recordable activities using this Log. However, if your committee is already the focal point of the majority of this type of information, then your existing minutes will probably reflect this already, and this Log will merely serve as a supplement to monitor and present activities.

Or, each department may be given this package and asked to begin to keep is own Log, and even go back in time to record issues that have been resolved that demonstrate good safety management. The departments can submit a copy quarterly or so to the safety committee or safety director for compilation. An on-site managers' workshop/seminar can be given on how to utilize this guide and log optimally. Please contact Elsie Tai for more information or to make arrangements.

Note: It is not advisable to provide safety committee minutes to an OSHA inspector. They would *not* normally ask to review them. However, if they do ask, it is better to sidestep the request as much as possible, by explaining that the minutes include information on residents that would be a breach of their privacy, etc. Ultimately, if an inspector insists on seeing the minutes, you may consider explaining in a cooperative and courteous manner, that you would really like to cooperate as much as possible, but for the issue of a breech of privacy of resident information and medical records, you respectfully request the inspector subpoena them. The problem is that those minutes will contain evidence of any inadequacies of your safety management. If you have identified hazards and have been unable or unwilling to correct them, it will show in these minutes. It is analogous to handing a loaded gun to an inspector, unless you are quite certain that there are no such instances of seemingly "willful negligence," or situations that could be construed as such within the minutes. The worst-case scenario is to have an item showing "pending for financial or budgeting reasons." OSHA interprets this as placing profits before employee safety. This could result in a willful violation, which increases the size of the penalty substantially. Nonetheless, insightful interviews with employees may reveal the same liability. In the end, despite potential citations, it will be the measure of prudent practices that employers will be held to.

3 OSHA Standards

I. INTRODUCTION

Requirements are rather specific and definitive when a standard does exist. The key standards that affect long-term care are OSHA 200 Log Recordkeeping; Bloodborne Pathogens; Hazard Communication; Lockout/Tagout; Emergency Plan: Fire Extinguisher Use; Emergency Eyewash; Machine Guarding; Electrical Safety; Medical and Exposure Records; Compressed Gas; and Slips, Trips, and Falls. Tuberculosis, Ergonomics, and Workplace Violence are also extremely relevant, if not vital to long-term care employee health and safety. However, OSHA does not have standards on these hazards at the time of this writing. OSHA reviews the management of these hazards, and citations may be given under the General Duty Clause. These nonstandard issues are covered in Chapter 4.

The following sections provide a comprehensive overview of each standard and how the requirements may be managed most effectively in a long-term care setting. Please refer to Chapter 1, Section III for guidance on determining penalties based on severity and probability of risk of injury for any of these hazards covered by these standards.

Table 3.1 is a list of all the cited standards from the previous year (October 1997 to September 1998). There may be more than one citation under a standard for each inspection. The penalties reflect current, more likely reduced amounts, not initial amounts. Keep in mind that since long-term care is relatively new ground for OSHA, these standards will shuffle around on the popularity list while inspectors find new or certain items easier to cite and others more difficult to defend.

Each state has its own top 20 list and the lists differ from the federal lists as they reflect the focus and interests of the local OSHA offices and perhaps some geographic differences in standards among facilities. The federal list is a nationwide reflection of larger trends in general, which may not impact an individual state, or area.

A risk map of the common types of health and safety hazards in a long-term care facility is presented in Appendix A. Keep in mind that one can be cited for both the existence of a physical hazard *and* the management deficiencies that allowed the hazard to develop, exist, and persist, especially if the hazard has been regulated by standard.

II. MANAGING ASBESTOS IN A FACILITY

A. INTRODUCTION

OSHA inspections of nursing homes do not normally include an asbestos inquiry. However, if renovations, demolition, construction, or questionable repair work is being done, or if there are signs it was done, or if employee interviews point to the issue, then asbestos will be investigated. If the building was built before 1980, then asbestos and materials potentially containing asbestos may be present in the facility. It is the responsibility of all building/facility owners to determine the existence and/or probability of the presence of asbestos and asbestos-containing materials and to manage them accordingly. Asbestos and asbestos-containing materials (ACM) generally do not pose a health hazard unless they are damaged or in a deteriorating condition that is likely to release fibers (friable). An asbestos management program evaluates the risks and manages them with regard to the logistics of the operation(s), sites of asbestos, type and use of the ACM, future plans for renovations, demolition, etc. The following is a synopsis to help readers

TABLE 3.1
Top 20 Most Frequently Cited OSHA Standards for Long-Term Care

No.	Standard	No. Cited	No. Inspected	Penalty, $	Description
1	Total	1150	247	479,672.00	
2	1910.1030	235	86	156,332.50	Bloodborne Pathogens
3	1910.305	102	63	28,899.50	Electrical, Wiring Methods, Components, and Equipment
4	1910.1200	79	50	11,722.50	Hazard Communication
5	1910.147	73	48	25,355.44	Lockout/Tagout—The Control of Hazardous Energy
6	1910.303	73	50	24,974.96	Electrical Systems Design, General Requirements
7	1910.212	65	54	44,527.80	Machines, General Requirements
8	1910.132	60	39	21,204.50	Personal Protective Equipment—General Requirements
9	1904.0002	56	55	11,180.00	OSHA 200 Log and Summary of Occupational Injuries and Illnesses
10	1910.0304	45	35	21,980.46	Electrical, Wiring Design and Protection
11	1910.151	43	43	31,830.75	Medical Services and First Aid
12	1910.0215	37	21	4,075.00	Abrasive Wheel Machinery
13	1910.0037	23	18	2,850.00	Means of Egress, General
14	1910.157	23	22	1,961.36	Portable Fire Extinguishers
15	1910.1020	20	22	0.00	Access to Medical and Exposure Records
16	1910.0022	15	13	2,125.00	Walking-Working Surfaces, General Requirements
17	1910.0133	15	15	4,868.75	Eye and Face Protection
18	1910.0146	15	7	7,477.83	Permit-Required Confined Spaces
19	1910.1001	15	7	15,252.50	Asbestos, Tremolite, Anthophyllite, and Actinolite
20	1910.0333	14	9	6,702.40	Electrical, Selection and Use of Work Practices
21	1910.0023	13	12	9,409.00	Guarding Floor and Wall Openings and Holes
22	1910.0138	12	12	2,804.00	Hand Protection
23	1910.0213	11	6	3,312.50	Woodworking Machinery Requirements
24	1910.0219	10	7	8,095.75	Mechanical Power-Transmission Apparatus
25	1910.0038	9	9	2,575.00	Employee Emergency Plans and Fire Prevention Plans
26	1904.0004	8	8	5,950.00	Supplementary Record (OSHA 101 or comparable A/I report)
27	1904.0005	8	8	0.00	Annual Summary, Occupational Injuries and Illnesses
28	1910.0106	8	7	2,682.00	Flammable and Combustible Liquids
29	1926.1101	8	3	5,070.00	Asbestos (Construction Standard, not General Industry Standard)
30	1910.0027	5	3	775.00	Fixed Ladders
31	1910.0334	5	5	550.00	Electrical, Use of Equipment
32	1910.0024	4	3	1,625.00	Fixed Industrial Stairs
33	1910.0134	4	4	0.00	Respiratory Protection
34	1910.0036	3	3	3,125.00	Means of Egress, General Requirements
35	1910.0101	3	3	2,500.00	Compressed Gases, General Requirements
36	1910.0145	3	3	0.00	Specifications, Accident Prevention Signs and Tags
37	5A1	2	2	1,850.00	General Duty Clause
38	1904.0006	2	2	0.00	Retention of Records
39	1910.0104	2	1	0.00	Oxygen
40	1910.0332	1	1	2,365.00	Electrical, Training
41	1910.165	1	1	975.00	Employee Fire Protection Alarm System
42	1910.0025	1	1	750.00	Portable Wood Ladders
43	1904.0007	1	1	500.00	Access to Records
44	1910.0265	1	1	437.50	Sawmills

Note: Also with one inspection, one citation, and no penalties were Hand and Portable Powered Tools and Equipment, General, Oxygen-Fuel Gas Welding and Cutting, Arc Welding and Cutting, Bakery Equipment, and Materials Handling.

gain some understanding of their risks, and responsibilities, and how they may best begin approaching the issue. *This is only a short synopsis. Consult your state and local authorities (DOH, DEP/DEC).*

B. HISTORY

Most of the well-documented health effects of asbestos exposure are based on long-term exposure to high levels of asbestos in certain industries, i.e., shipbuilding, mining, milling, and fabricating. Symptoms of disease do not appear until perhaps 20 years or more after exposure. The three specific diseases with which asbestos exposure has been linked are asbestosis (fibrous scarring of the lungs), lung cancer, and mesothelioma (a cancer of the lining of the chest or abdominal cavity). Average airborne asbestos levels in buildings seem to be very low. Correspondingly, the health risk to most building occupants also appears to be very low. The mere presence of asbestos in a building is not cause for alarm, or immediate removal, as removal is often *not* the best course of action to reduce asbestos exposure. Improper removal can *create* a dangerous situation where none previously existed. Asbestos removal tends to elevate the airborne level of asbestos fibers unless all safeguards are properly applied. Asbestos removal is only required by the EPA before major building demotion or renovation activities to prevent significant public exposure to airborne asbestos fibers during those highly disturbing activities. It may be more effective to manage existing asbestos in place.

Asbestos was popularly used in commercial product because of its strength, inflammability, corrosion resistant, and insulating properties. In the United States, commercial use began in the early 1900s and peaked in the period between World War II and the 1970s. Under the EPA Clean Air Act of 1970, ACM have been regulated. OSHA also has specific rules on exposure levels, labeling, employee training, personal protective equipment, respiratory protection standards, and work practices regarding employees who may be exposed. The EPA banned several major types of asbestos materials, such as spray-applied insulation, fireproofing, and acoustical surfacing materials in the mid-1970s because of the growing concern of the health effects. The EPA Asbestos Ban and Phasedown Rule was promulgated in July 1989, which applies to new product manufacture, importation, and processing. It essentially banned almost all asbestos-containing products in the United States by 1997. This rule does not require removal of asbestos currently in place in buildings.

C. MOST LIKELY SCENARIO

Most long-term care facilities will likely *manage asbestos in place.* An asbestos management program with an operations and maintenance (O&M) program to manage the in-place asbestos should be developed and implemented. This requires an initial assessment of the presence, extent, and risk inherent in the ACM in the building. Most OSHA requirements do not apply until employees may be reasonably expected to be exposed to the permissible exposure limit (PEL). This usually occurs during renovations and construction. The only exception is the *OSHA requirement that maintenance and custodial staff receive annual Asbestos Awareness Training if they work near or around ACM or presumed ACM (PACM).* A facility should have had an asbestos survey to determine where, if at all, ACM are present. If not present, then analysis should be done on materials before certain repairs are initiated (i.e., breaking a wall to fix a leaking pipe). Most maintenance activities can be conducted by in-house staff unless their risk of contact with ACM is elevated as a result of potential disturbance to the ACM. The bulk of this guide is dedicated to this scenario. It provides basic guidelines for an asbestos management plan with a high-quality O&M program.

In the instance where ACM may be disturbed, and staff may be exposed to asbestos at or near the PEL, consider having a contractor encapsulate, enclose, seal, or contain the materials in an acceptable manner so that contact for custodial and maintenance staff would be *unlikely,* as if the materials were in good condition. Asbestos removal is costly, and can pose a greater health hazard if improperly executed. If management in place is possible and reasonable, it would be the preferred method. Of course, the decision should be predicated on the practical logistics of the site, the actual material(s) in question, the work area, public access, size of the material, size of the affected area, ability to isolate, relating conditions, and future work plans, i.e., water damage, repairs, emergency repairs, renovations, demolition, etc.

D. ANOTHER POSSIBLE SCENARIO

Since many long-term care facilities appear to be engaged in renovations, and some demolitions, they may be engaged in asbestos removal/abatement, *as the asbestos must be removed before such activities commence.* Again, renovations, construction, work requiring a permit, and even certain maintenance repair work may attract the attention of OSHA inspectors to this issue. The rule of thumb should basically be that no long-term care facility employee engage in any of these activities. Long-term care employees should only be allowed to work near and around ACM if the likelihood of disturbance is very low, if "contact is unlikely," and if the material is intact and in good condition. If the likelihood of material disturbance and employee contact is elevated to "possible accidental disturbance" or "intentional or likely disturbance," then consider the use of qualified contractors.

It is usually not worth the investment and administrative effort to qualify, train, and manage the programs properly in order that in-house employees can handle asbestos, unless the facility or campus is large and the asbestos presence is serious. This also involves medical monitoring, extensive accredited training, a respiratory protection program, hazardous wastes management, air monitoring, etc. (*Refer to state and local agencies for specific requirements.*) The O&M program can also be used to minimize exposure in the higher-risk categories. The nature of the program would be different, as a reflection of the higher risks involved.

There may be considerations to engage in an abatement project. For example, removal is required by NESHAP (National Emissions Standards for Hazardous Air Pollutants) regulations for projects that would break up more than a specified minimum amount of ACM; specifically, at least 160 ft^2 of surfacing or miscellaneous material, or at least 260 linear ft of thermal system insulation. Check also with state and local health and environmental authorities for further clarifications and additional rules that may apply. Often, other extenuating circumstances can cause the building owner to engage in abatement, i.e., difficulty obtaining insurance or financing real estate transactions. Again, these decisions are predicated on the practical logistics of the site, the actual material(s) in question, the work area, public access, size of the material, size of the affected area, ability to isolate, relating conditions, and future work plans, etc., along with other prerequisites.

In selecting a consultant, ensure that the consultant has a background in engineering, architecture, industrial hygiene, safety, or a similar field. Registered and/or board-certified experts are recommended. Ensure that no "conflict of interest" exists. Consultants should be affiliated neither with the abatement contractors who may be used on a recommended abatement project nor with analytical laboratories that perform sampling analyses. Interview several consultants and check their references. Contracts with service trades or abatement companies should include the following provisions to ensure that the service or abatement workers can and will follow appropriate work practices:

- Proof that the contractor's workers have been properly notified about the ACM in the building and that they are properly trained and accredited (as necessary) to work with ACM.
- Copies of respiratory protection, medical surveillance, and worker training documentation required by OSHA, EPA, and state and local authorities.
- Notification to building tenants and visitors that abatement activity is under way.
- Written work practices that must be submitted by the vendor/contractor for approval or modification by the in-house Asbestos Program Manager. The vendor/contractor should then agree to abide by the work practices as finally accepted by the manager.
- Assurance that the contractor will use proper work area isolation techniques, proper equipment, and sound waste-disposal practices.

- Historical air-monitoring data for representative examples of the contractor's previous projects, with emphasis on projects similar to those likely to be encountered.
- Provisions for inspections of the area by the owner's representative to ensure that the area is acceptable for reentry of occupants/tenants.
- A resume for each abatement contractor/supervisor or maintenance crew chief, known as the "competent person" in the OSHA standard and the EPA Worker Protection Rule.
- Criteria to be used for determining successful completion of the work (i.e., visual inspections and air-monitoring results).
- Any other information deemed necessary by legal counsel.
- Notification to the EPA/other appropriate agencies if the abatement project is sufficiently large.

E. SOME RELEVANT DEFINITIONS

Air Plenum: Space used to convey air in building/structure (i.e., above suspended ceiling).

Asbestos-containing material (ACM): Any material containing more than 1% asbestos.

Building/facility owner: The legal entity, including a lessee, that exercises control over management and record-keeping functions relating to a building and/or facility's asbestos information and management.

Friable asbestos: Any materials that contain more than 1% asbestos, and that can be crumbled, pulverized, or reduced to powder by hand pressure. These may also include previously nonfriable material that becomes broken or damaged by mechanical force.

Presumed asbestos containing material (PACM)—Includes thermal system insulation and surfacing material found in *buildings constructed no later than 1980*. The designation of materials as PACM may be rebutted if the employer and/or building owner can demonstrate that the PACM does not contain asbestos. (The information, data, and analysis supporting the determination that PACM does not contain asbestos shall be strictly and accurately maintained and passed onto subsequent responsible parties.) This can be achieved by:

- Completing an AHERA (Asbestos Hazard Emergency Response Act, 40 CFR 763, Subpart E) inspection that demonstrates that no ACM is present in the material, or
- Performing a test of the materials containing PACM, which demonstrates that no ACM is, in fact, present in the material. This includes analysis of bulk samples collected as per 40 CFR 763.86 protocols. (The tests, evaluation, and sample collection shall be conducted by an accredited inspector or by a certified industrial hygienist. The analysis of the sample must be performed by a laboratory with proficiency demonstrated by current successful participation in a nationally recognized testing program.)
- One may also demonstrate that flooring material, including associated mastic and backing, does not contain asbestos, by a determination of an industrial hygienist based upon recognized analytical techniques showing that the material is not ACM.

Phase contrast microscopy (PCM): Tool commonly used for personal air sample analysis and as a screening tool for area monitoring. It costs about $25 and is adequate as an initial screening tool to confirm or discount the presence of ACM.

Surfacing ACM: Surfacing material that contains more than 1% asbestos.

Surfacing material: Material that is sprayed, troweled-on, or otherwise applied to surfaces (such as acoustical plaster on ceilings and fireproofing materials on structural members, or other materials on surfaces for acoustical, fireproofing, and other purposes).

Thermal system insulation (TSI): ACM applied to pipes, fittings, boilers, breeching, tanks, ducts, or other structural components to prevent heat loss or gain, thermal system insulation that contains more than 1% asbestos.

Transmission electron microscope (TEM): The most accurate and the preferred method of analysis. Air samples (if any) collected under an O&M program, would require its use. However, it costs about ten times more than the PCM test.

It would be best to operate within the confines of the following scenarios

- Valid verification and documentation that no ACM is present. Maintain these as permanent records and pass on to subsequent owners (share data with tenants).
- Verify that ACM is indeed present. Confirm that it is in good condition and poses very low risk. Employees are *unlikely* to come on contact with it. This is initiated as part of the asbestos management plan. Design and implement an O&M program based on the information provided.
- Assume materials are PACM (if building built before 1980). Confirm that they are in good condition and pose very low risk. Employees are *unlikely* to come into contact with them. This is initiated as part of the asbestos management plan. Design and implement an O&M program based on this information.
- If the ACM or PACM is in deteriorating condition, where there is "possible accidental disturbance" or "intentional or likely disturbance," then find professional, qualified assistance in helping decide the best course of action, including contractors to encapsulate, seal, contain, or remove the material properly.
- If all ACM is removed, then no program is necessary in the future. Maintain records.
- If some ACM remains, then the asbestos management program should continue and the O&M plan must be continually followed.

Note: Do not provide custodial or maintenance employees with disposable paper dust masks as an added measure of protection, even around intact ACM or PACM in good condition. OSHA, EPA, and NIOSH are on record as not recommending single use, disposable paper dust masks for use against asbestos; in fact, OSHA has *disallowed* their use against airborne asbestos fibers.

F. Critical Points Summary

- Many of the requirements are applicable when employees are, or *may reasonably be expected to be,* exposed to airborne concentrations *at or above the permissible exposure limits.*
- Certain materials defined above will be presumed to be ACM, or are PACM, if the building was built before 1980, even without verification that this is, in fact, true.
- Valid verification and documentation that PACM is not ACM relieves building owners from the requirements of those with ACM or PACM.
- Positive identification of asbestos requires laboratory analysis; information labels or visual examination are not sufficient.

G. Asbestos Management

An asbestos management program should be administered by a qualified manager with authority to oversee and direct custodial/maintenance staff and contractors with regard to all asbestos-related activities:

- The program is initiated by a building inspection by a trained, qualified, experienced inspector to locate and assess the condition of all ACM or PACM in the building.
- The inspection results serve as the basis for establishing an O&M program. O&M procedures may not be sufficient for certain ACM that is significantly damaged or in highly accessible areas.

- The written O&M programs are periodically reviewed.
- Alternatives or control options that may be implemented under an O&M program include repair, encapsulation, enclosure, encasement, and minor removal.

A basic O&M program contains the following:

- A notification program* to inform building occupants, workers, and tenants about the location of ACM and how to avoid disturbing the ACM or PACM.
- Distribution of written notices or posting of signs or labels in central locations where affected occupants can see them, holding awareness/information sessions (depends on the type and location of the ACM and number of people affected).
- Use of "Caution—Asbestos—Do Not Disturb" or "Danger—Contains Asbestos Fibers—Avoid Creating Dust" signs in service and maintenance areas (boiler rooms) directly adjacent to thermal system insulation to alert and remind workers not to disturb the ACM inadvertently.
- Furnishing of all boilers, pipes, and other equipment with ACM or PACM in service areas where damage may occur with prominent warning signs affixed or placed next to the ACM/PACM. As an alternative, color coding may be used to identify ACM in certain situations provided that all potentially involved parties understand the system.

It may be necessary to make special provisions for workers who cannot read English (clear verbal information and signs, foreign language interpretations, pictorials, graphs, etc.).

1. Labeling

The information provided should contain the following to the extent they reflect building conditions:

- ACM has been found in the building and is located in areas where the material could be disturbed.
- The condition of the ACM, and the response that is appropriate for that condition, should be stated.
- Asbestos only presents a health hazard when fibers become airborne and are inhaled. The mere presence of ACM does not represent a health hazard.
- The ACM is found in the following location(s): _____
- Do not disturb the ACM (do not push furniture against or damage thermal system insulation, etc.).
- Report any evidence of disturbance or damage of ACM to [name, location, and phone number of asbestos program manager].
- Report any dust/debris that might come from ACM or suspect ACM, any change in the condition of the ACM, or any improper action relative to ACM of building personnel to [name, location, phone number of asbestos program manager].
- Cleaning and maintenance personnel are taking special precautions during their work to clean up any asbestos debris properly to guard against disturbing ACM.
- All ACM is inspected periodically and additional measures will be taken if necessary.

*Note: Labels are not required where asbestos fibers have been modified by a bonding agent, coating, binder, or other material provided that the manufacturer can demonstrate that during any reasonably foreseeable use, handling, storage, disposal, processing, or transportation, no airborne concentrations of fibers of asbestos in excess of the permissible exposure levels will be released, or asbestos is present in a product in concentrations less than 1%.

2. Periodic Surveillance and Reinspection

- Maintain periodic surveillance and reinspection of ACM or PACM at regular intervals by trained workers or properly trained inspectors.
- Visually reinspect all ACM at regular intervals (every 6 months).
- Ensure any ACM damage or deterioration will be detected and corrective action taken through ongoing reports of changes in the condition of ACM made by service workers along with the reinspection.
- Determine relative degree of damage and assess the likelihood of future fiber release. (Maintain a set of visual records, photographs, or videotapes.)
- Visually inspect to recognize situations and anticipate future exposure (i.e., worsening water damage).
- Include a sampling strategy* to supplement visual/physical evaluation. This is generally not recommended as it is costly and requires qualified oversight.
- Within the scope of an O&M, analyze the samples using the TEM method.
- Remember that air monitoring can only detect a problem after it has occurred and fibers have been released.
- Consider abatement alternatives.

3. Work Control/Permit System

A work control/permit system is a system to control work that could disturb ACM or PACM. This system requires the person requesting work to submit a job request form to the asbestos program manager and obtain authorization, before any work is done. Upon receipt of a prework job request form, the asbestos program manager should:

- Refer to written records, building plans and specifications, and any building ACM inspection reports to determine whether ACM is present in the area where work will occur. If ACM is present, but it is not anticipated that the material will be disturbed, then the presence of ACM should be noted on the permit form and additional instruction provided on the importance of not disturbing the ACM.
- If ACM is both present and *likely* to be disturbed, then the site should be visited by a qualified person to determine what work practices should be instituted to minimize the release of asbestos fibers during maintenance activity.

This determination should be recorded on the maintenance work authorization form, which is then forwarded to the in-house maintenance supervisor, or the contractor requesting the authorization. The asbestos program manager should ensure that a copy of both the request and the authorization forms (if granted) are placed in the permanent files.

Where the task is not covered by previously approved standard work practices, the asbestos program manager should make sure that the appropriate work practices and protective measures are used for the job. For all jobs where contact with ACM is likely, the qualified person or manager should visit the work site when the work begins to see that the job is being performed properly. For lengthy jobs where disturbance of ACM is intended or likely, periodic inspections should be made for the duration of the project.**

---

*Be careful not to misinterpret findings, as collection samples may be misleadingly low compared with an episodic event such as repair work or accidental disturbance. Or results may be misleadingly high if overly aggressive sampling methods are used, not enough samples are collected, or sampling occurred during an episodic event, which is not representative of routine exposure.

**Note: The last three steps would normally not apply to routine maintenance work and would more likely involve contractor work where a qualified person should probably be hired to execute these tasks as part of the scope of the contract/project.

Observations should be provided on an evaluation of work form, any deviation from standard and approved work practices should be recorded immediately on this form and the practices should be immediately corrected and reported.* Upon completion of the work, a copy of the evaluation form should be placed in the permanent asbestos file for the building.*

4. Operations and Maintenance Work Practices

O&M work practices are defined to avoid or minimize fiber release during activities affecting ACM or PACM; a summary of when to apply various work practice is presented in Table 3.2. This section should focus on special sets of work practices for custodial, maintenance, and construction staff. The nature and extent should be tailored to the likelihood that the ACM will be disturbed and that fibers will be released. In general, four broad categories of O&M work practices are recognized.

Worker Protection Program: A program that helps to ensure custodial and maintenance staff are adequately protected from asbestos exposure. (If respirators are provided, a working written Respiratory Protection Program with fit testing, annual training, and medical clearance, and proper selection of equipment is necessary.) [*Note:* This includes the provision in the OSHA Asbestos Standard requiring employers to provide employees who perform housekeeping operations in an area that contains ACM or PACM with an asbestos awareness training course, annually, which shall at a minimum contain health effects of asbestos locations of ACM and PACM in the building; recognition of ACM and PACM damage and deterioration, and requirements in the standard relating to housekeeping and proper response to fiber release episodes.]

TABLE 3.2
Summary of When to Apply Key O&M Work Practices

	Likelihood of ACM Disturbance		
	Contact Unlikely	Accidental Disturbance Possible	Disturbance Intended/ Likely
Management Responsibilities			
Need prework approval from asbestos program manager	Review by program manager	Yes	Yes
Special scheduling or access control	No	Yes	Yes
Supervision needed	No	Initial, at least	Yes
HVAC system modification (in area where work takes place)	None	As needed	Shut down
Area containment	None	Drop cloths, minienclosures	Yes (full or partial containment)
Personal Protective Equipment			
Respiratory protection*	Available for use	Yes	Yes
Protective clothing	None	Review by asbestos program manager	Yes
Work Practices			
Use of wet methods	No	As needed	Yes
Use of HEPA (high efficiency particulate air) vacuum*	Available for use	Available for use	As needed

* Involves other requirements.

Note: The last three steps would normally not apply to routine maintenance work and would more likely involve contractor work where a qualified person should probably be hired to execute these tasks as part of the scope of the contract/project.

Basic O&M Procedures: Basic procedures used to perform routine custodial and maintenance tasks that may involve ACM. [*Note:* This includes recent OSHA Asbestos Standard amendments on the care of asbestos-containing flooring materials:

- Sanding of asbestos-containing floor material is prohibited.
- Stripping of finishes shall be conducted using low abrasion pads at speeds lower than 300 rpm and wet methods.
- Burnishing or dry buffing may be performed only on asbestos-containing flooring that has sufficient finish so that the pad cannot contact the ACM.]

Special O&M Cleaning Techniques: Clean up asbestos fiber on a routine basis (at which point one may want to test, and/or encapsulate, or remove it so this is *not* a routine since such tasks may arguably expose staff to levels reaching regulated action levels and permissible exposure limits, requiring a great number of other elements).

Procedures for Asbestos Fiber Release Episode: Necessary if moderate to large amounts of ACM are disturbed (contractor work).

Basic O&M procedures fall into three categories of maintenance activities and their potential for disturbing ACM:

- Those that are unlikely to involve any direct disturbance of ACM (i.e., cleaning shelves or countertops with a damp cloth).
- Those that may cause accidental disturbance of ACM (working on a fixture near a ceiling with surfacing ACM, accidentally damaging ACM with broom handles or ladders while performing other tasks).
- Those that involve intentional, small-scale manipulation or disturbance of ACM (removing small segment of thermal system insulation or ACM to repair a pipe leak).

5. Record Keeping

- Employee exposure, sample analysis results, consultant's reports, etc. should be maintained in the records.
- All asbestos-sampling reports, inspection reports, consultant's reports must be maintained in a permanent file and the information passed on to subsequent owners.
- Objective data used to qualify for exemptions from OSHA initial monitoring requirements must be maintained for the duration of the exemption.

If a facility is in any asbestos removal, or work beyond the risk category of *contact unlikely*, into the risk categories of *possible accidental disturbance* or *intentional or likely disturbance,* then generation of the following documents, which should be kept permanently and, in the case of employee records, must be kept for at least 30 years after the last day of employment, is necessary.

- Medical records of employees subject to the medical surveillance program
- All employee training records for 1 year beyond the last date of employment
- Personal air sampling records

H. CONCLUSION

This is only a short synopsis to prime the beginner. Seek qualified professional assistance to determine the presence of asbestos in a facility, and design a plan of action and an ongoing maintenance plan, as necessary. Refer to the end of this section (page 67) for a list of references.

Following are sample forms mentioned in this section.

SAMPLE FORM: REINSPECTION OF ACM

Location of asbestos-containing material [address, building, room, or general description]:

_____

_____

_____

Type of asbestos-containing material(s):

Sprayed or troweled on ceilings or walls
Sprayed or troweled on structural members
Insulation on pipes, tanks, or boiler
Other (describe):

_____

_____

Abatement Status:
The material has been encapsulated_____, enclosed_____, neither_____, removed_____.
Assessment:

1. Evidence of physical damage:

_____

2. Evidence of water damage:

_____

3. Evidence of delamination (separation of one layer from another) or other damage:

_____

4. Degree of accessibility of the material:

_____

5. Degree of activity near the material:

_____

6. Location in an air plenum, air shaft, or airstream:

_____

7. Other observations (including the condition of the encapsulant or enclosure, if any):

_____

Recommended Action:

_____

_____

Signed:_____ Date:_____/_____/_____
    [evaluator]

SAMPLE FORM: JOB REQUEST FOR MAINTENANCE WORK

Name:_____ Date:_____/_____/_____
Telephone No.:_____ Job Request No.:_____
Request starting date:_____ Anticipated finish date:_____/_____/_____
Address, building, and room number (or description of area) where work is to be performed:

_____

_____

_____

Description of work:

_____

_____

_____

_____

_____

Description of any asbestos-containing material that might be affected, if known (include location and type):

_____

_____

_____

_____

_____

Name and telephone number of requestor:

_____

Name and telephone number of supervisor:

_____

Submit this application to:

_____

[Asbestos Program Manager]
Note: An application must be submitted for all maintenance work whether or not asbestos-containing material might be affected. An authorization must then be received before any work can proceed.

_____ Granted (Job Request No. _____)
_____ With conditions*
_____ Denied
*Conditions: _____

_____

SAMPLE FORM: MAINTENANCE WORK AUTHORIZATION

No._____
Authorization
Authorization is given to proceed with the following maintenance work:

_____

_____

_____

_____

Presence of Asbestos-Containing Materials
_____ Asbestos-containing materials are not present in the vicinity of the maintenance work.
_____ ACM is present, but its disturbance is not anticipated; however, if conditions change, the asbestos program manager will reevaluate the work request prior to proceeding.
_____ ACM is present, and may be disturbed.
Work Practices if Asbestos-Containing Materials Are Present
The following work practices shall be employed to avoid or minimize disturbing asbestos:

_____

_____

_____

_____

_____

Personal Protection if Asbestos-Containing Materials Are Present The following equipment/
clothing shall be used/worn during the work to protect workers [at this point, qualified expert assistance should be consulted]:

_____

_____

_____

_____

_____

Special Practices and/or Equipment Required [consult qualified expert]:

_____

_____

_____

_____

_____

Signed: _____ Date: _____/_____/_____
[Asbestos Program Manager]

REFERENCES

Contact the EPA TSCA Assistance Hotline, 202-554-1404 for more information and any more recent titles.

EPA Guidance for Controlling Asbestos-Containing Materials in Building (also known as the *Purple Book*),
     EPA Publ. 560/5/85/024.

EPA Guidance for Service and Maintenance Personnel, EPA Publ. 560-5-85-018.

EPA Abatement of Asbestos-Containing Pipe Insulation Tech. Bull. 1986-2.

EPA Managing Asbestos in Place—A Building Owner's Guide to Operations and Maintenance Programs for
     Asbestos-Containing Materials, 20T-2003 (TS-799), July 1990.

EPA Operations and Maintenance Work Practices Manual, 1991.

Keyes, D and Chesson, J, A Guide to Monitoring Airborne Asbestos in Buildings, Environmental Sciences, Inc.,
     105 E. Speedway Blvd., Tucson, Arizona 85705, 1989.

OSHA Respiratory Protection Standard 29 CFR 1910.134.

U.S. Department of Labor, OSHA Regulations 29 CFR 1910.1001, General Industry Asbestos Standard, and
     1926.58 Construction Industry Asbestos Standard.

U.S. EPA, National Emission Standards for Hazardous Air Pollutants, (NESHAP) Regulations 40 CFR 61,
     April 5, 1984.

U.S. EPA, Asbestos in Buildings: Simplified Sampling Scheme for Surfacing Materials (also known as the *Pink
     Book*), Publ. 560/5-85-030A, 1985.

U.S. EPA, Measuring Airborne Asbestos Following an Abatement Action (also known as the *Silver Book*), Publ.
     600/4/85/049, 1985.

U.S. EPA, Asbestos Ban and Phaseout Rule, 40 CFR 763.160 to 763.179, *Fed. Regis.*, July 12, 1989.

III.   EMPLOYEE ACCESS TO EXPOSURE AND MEDICAL RECORDS

A.   GOAL

This standard allows employees and former employees, or their designated representatives, access
to their accident reports, physicians' exams, MSDSs, any other comprehensive reports (i.e., accident
analysis), which include the individual's case in the data collected and reviewed, etc.

B.   SUMMARY

Employers must make employees aware of their right to access these types of records, and that the
records must be made available to them within 15 days of their request, *at no cost to the employee*,
within reasonable means (i.e., accessible location and time). There are no specific requirements how

this should be done. Employers are also required to retain a copy of the standard on site, usually in an office, for employee or inspector review. This accessibility allows employees and their representatives to identify employers, and past employers, that may be potentially liable for an employee's chronic or acute illness.

C. UPDATE

Originally, this standard was promulgated to establish the rights of employees to access medical and health information related to their chemical exposures. Although this still holds true, the focus of this accessibility has been expanded and highlighted to include nonchemical exposure issues, like bloodborne pathogens and tuberculosis in the health care setting.

Inspections do not generally look for documentation of compliance to this standard. However, inquiries will arise if employee interviews point to deficiencies in this area.

D. CONTENTS

A sample form letter that can be either posted or enclosed in payroll as an envelope stuffer is provided in Section III.E. Review the letter carefully to make sure that the administration understands the terms disclosed, and that management can adopt them into policy. Then fill in the blanks appropriately and post or distribute. It is generally accepted that this should be done upon hire and annually thereafter.

A copy of the Standard, which must be kept in the administrative office for employee, or more likely, inspector review, is provided on page 69.

E. SAMPLE LETTER

TO: All Employees

SUBJECT: Access to Employee Exposure and Medical Records

In an effort to comply with Occupational Safety & Health regulations, this will serve to inform you of your rights as employees to access any exposure and medical records that may exist in your file. Exposure Records may include, but are not limited to, information on workplace monitoring data; biological monitoring results; records on biohazard exposure incidences; Material Safety Data Sheets; and any other record that identifies toxic substances and harmful physical agents; etc.

Medical Records may include, but are not limited to, information on any medical examinations and laboratory tests; medical opinions, diagnosis, progress notes, and recommendations; descriptions of treatments and prescriptions; and medical complaints; etc.

Your rights as employees are as follows under the OSHA regulation titled, "Access to Employee Exposure and Medical Records":

1. You or a "designated representative" have a right to review your exposure and medical records. Access to your records will be provided in a reasonable amount of time, but no later than 15 days after your request.

2. A copy of your records can be provided to you free of charge. We will either loan you the records to copy or we will allow you to copy them on the office photocopier at our facility.

3. If your records have been used to develop analytical reports, then you have a right to those reports. No other employee or agents may have access to your medical records without your explicit written consent.

Your exposure and medical records are kept in the _____ Department. All inquiries regarding this issue should be referred to _____ during normal business hours.

A copy of the "Access to Employee Exposure and Medical Records" OSHA Standard is also available for your review in this office.

_____

[signed by an officer]

F. REGULATIONS

1910.1020 Title: Access to Employee Exposure and Medical Records

(Subpart Z Subpart Title Toxic and Hazardous Substances)

1910.1020(a)

(a) "Purpose." The purpose of this section is to provide employees and their designated representatives a right of access to relevant exposure and medical records; and to provide representatives of the Assistant Secretary a right of access to these records in order to fulfill responsibilities under the Occupational Safety and Health Act. Access by employees, their representatives, and the Assistant Secretary is necessary to yield both direct and indirect improvements in the detection, treatment, and prevention of occupational disease. Each employer is responsible for assuring compliance with this section, but the activities involved in complying with the access to medical records provisions can be carried out, on behalf of the employer, by the physician or other health care personnel in charge of employee medical records. Except as expressly provided, nothing in this section is intended to affect existing legal and ethical obligations concerning the maintenance and confidentiality of employee medical information, the duty to disclose information to a patient/employee or any other aspect of the medical-care relationship, or affect existing legal obligations concerning the protection of trade secret information.

1910.1020(b)(1)

(b) "Scope and application." (1) This section applies to each general industry, maritime, and construction employer who makes, maintains, contracts for, or has access to employee exposure or medical records, or analyses thereof, pertaining to employees exposed to toxic substances or harmful physical agents.

1910.1020(b)(2)

(2) This section applies to all employee exposure and medical records, and analyses thereof, of such employees, whether or not the records are mandated by specific occupational safety and health standards.

1910.1020(b)(3)

(3) This section applies to all employee exposure and medical records, and analyses thereof, made or maintained in any manner, including on an in-house or contractual (e.g., fee-for-service) basis. Each employer shall assure that the preservation and access requirements of this section are complied with regardless of the manner in which records are made or maintained.

1910.1020(c)(1)

(c) "Definitions." (1) "Access" means the right and opportunity to examine and copy.

1910.1020(c)(2)

(2) "Analysis using exposure or medical records" means any compilation of data or any statistical study based at least in part on information collected from individual employee exposure or medical records or information collected from health insurance claims records, provided that either the analysis has been reported to the employer or no further work is currently being done by the person responsible for preparing the analysis.

1910.1020(c)(3)

(3) "Designated representative" means any individual or organization to whom an employee gives written authorization to exercise a right of access. For the purposes of access to employee exposure

records and analyses using exposure or medical records, a recognized or certified collective bargaining agent shall be treated automatically as a designated representative without regard to written employee authorization.

1910.1020(c)(4)

(4) "Employee" means a current employee, a former employee, or an employee being assigned or transferred to work where there will be exposure to toxic substances or harmful physical agents. In the case of a deceased or legally incapacitated employee, the employee's legal representative may directly exercise all the employee's rights under this section.

1910.1020(c)(5)

(5) "Employee exposure record" means a record containing any of the following kinds of information:

1910.1020(c)(5)(i)

(i) Environmental (workplace) monitoring or measuring of a toxic substance or harmful physical agent, including personal, area, grab, wipe, or other form of sampling, as well as related collection and analytical methodologies, calculations, and other background data relevant to interpretation of the results obtained;

1910.1020(c)(5)(ii)

(ii) Biological monitoring results which directly assess the absorption of a toxic substance or harmful physical agent by body systems (e.g., the level of a chemical in the blood, urine, breath, hair, fingernails, etc.) but not including results which assess the biological effect of a substance or agent or which assess an employee's use of alcohol or drugs;

1910.1020(c)(5)(iii)

(iii) Material safety data sheets indicating that the material may pose a hazard to human health; or

1910.1020(c)(5)(iv)

(iv) In the absence of the above, a chemical inventory or any other record which reveals where and when used and the identity (e.g., chemical, common, or trade name) of a toxic substance or harmful physical agent.

1910.1020(c)(6)(i)

(6)(i) "Employee medical record" means a record concerning the health status of an employee which is made or maintained by a physician, nurse, or other health care personnel, or technician, including:

1910.1020(c)(6)(i)(A)

(A) Medical and employment questionnaires or histories (including job description and occupational exposures),

1910.1020(c)(6)(i)(B)

(B) The results of medical examinations (pre-employment, pre-assignment, periodic, or episodic) and laboratory tests (including chest and other X-ray examinations taken for the purpose of establishing a base-line or detecting occupational illnesses and all biological monitoring not defined as an "employee exposure record"),

1910.1020(c)(6)(i)(C)

(C) Medical opinions, diagnoses, progress notes, and recommendations,

1910.1020(c)(6)(i)(D)

(D) First aid records,

1910.1020(c)(6)(i)(E)

(E) Descriptions of treatments and prescriptions, and

1910.1020(c)(6)(i)(F)

(F) Employee medical complaints.

1910.1020(c)(6)(ii)

(ii) "Employee medical record" does not include medical information in the form of:

1910.1020(c)(6)(ii)(A)

(A) Physical specimens (e.g., blood or urine samples) which are routinely discarded as a part of normal medical practice, or

1910.1020(c)(6)(ii)(B)

(B) Records concerning health insurance claims if maintained separately from the employer's medical program and its records, and not accessible to the employer by employee name or other direct personal identifier (e.g., social security number, payroll number, etc.), or

1910.1020(c)(6)(ii)(C)

(C) Records created solely in preparation for litigation which are privileged from discovery under the applicable rules of procedure or evidence; or

1910.1020(c)(6)(ii)(D)

(D) Records concerning voluntary employee assistance programs (alcohol, drug abuse, or personal counseling programs) if maintained separately from the employer's medical program and its records.

1910.1020(c)(7)

(7) "Employer" means a current employer, a former employer, or a successor employer.

1910.1020(c)(8)

(8) "Exposure" or "exposed" means that an employee is subjected to a toxic substance or harmful physical agent in the course of employment through any route of entry (inhalation, ingestion, skin contact, or absorption, etc.), and includes past exposure and potential (e.g., accidental or possible) exposure, but does not include situations where the employer can demonstrate that the toxic substance or harmful physical agent is not used, handled, stored, generated, or present in the work-place in any manner different from typical non-occupational situations.

1910.1020(c)(9)

(9) " Health professional" means a physician, occupational health nurse, industrial hygienist, toxicologist, or epidemiologist, providing medical or other occupational health services to exposed employees.

1910.1020(c)(10)

(10) "Record" means any item, collection, or grouping of information regardless of the form or process by which it is maintained (e.g., paper document, microfiche, microfilm, X-ray film, or automated data processing).

1910.1020(c)(11)

(11) "Specific chemical identity" means a chemical name, Chemical Abstracts Service (CAS) Registry Number, or any other information that reveals the precise chemical designation of the substance.

1910.1020(c)(12)(i)

(12)(i) "Specific written consent" means a written authorization containing the following:

1910.1020(c)(12)(i)(A)

(A) The name and signature of the employee authorizing the release of medical information,

1910.1020(c)(12)(i)(B)

(B) The date of the written authorization,

1910.1020(c)(12)(i)(C)

(C) The name of the individual or organization that is authorized to release the medical information,

1910.1020(c)(12)(i)(D)

(D) The name of the designated representative (individual or organization) that is authorized to receive the released information,

1910.1020(c)(12)(i)(E)

(E) A general description of the medical information that is authorized to be released,

1910.1020(c)(12)(i)(F)

(F) A general description of the purpose for the release of the medical information, and

1910.1020(c)(12)(i)(G)

(G) A date or condition upon which the written authorization will expire (if less than one year).

1910.1020(c)(12)(ii)

(ii) A written authorization does not operate to authorize the release of medical information not in existence on the date of written authorization, unless the release of future information is expressly authorized, and does not operate for more than one year from the date of written authorization.

1910.1020(c)(12)(iii)

(iii) A written authorization may be revoked in writing prospectively at any time.

1910.1020(c)(13)

(13) "Toxic substance or harmful physical agent" means any chemical substance, biological agent (bacteria, virus, fungus, etc.), or physical stress (noise, heat, cold, vibration, repetitive motion, ionizing and non-ionizing radiation, hypo- or hyperbaric pressure, etc.) which:

*(i) Is listed in the latest printed edition of the National Institute for Occupational Safety and Health (NIOSH) Registry of Toxic Effects of Chemical Substances (RTECS) which is incorporated by reference as specified in Sec. 1910.6; or

1910.1020(c)(13)(ii)

(ii) Has yielded positive evidence of an acute or chronic health hazard in testing conducted by, or known to, the employer; or

1910.1020(c)(13)(iii)

(iii) Is the subject of a material safety data sheet kept by or known to the employer indicating that the material may pose a hazard to human health.

1910.1020(c)(14)

(14) "Trade secret" means any confidential formula, pattern, process, device, or information or compilation of information that is used in an employer's business and that gives the employer an opportunity to obtain an advantage over competitors who do not know or use it.

1910.1020(d)(1)

(d) "Preservation of records." (1) Unless a specific occupational safety and health standard provides a different period of time, each employer shall assure the preservation and retention of records as follows:

1910.1020(d)(1)(i)

(i) "Employee medical records." The medical record for each employee shall be preserved and maintained for at least the duration of employment plus thirty (30) years, except that the following types of records need not be retained for any specified period:

1910.1020(d)(1)(i)(A)

(A) Health insurance claims records maintained separately from the employer's medical program and its records,

1910.1020(d)(1)(i)(B)

(B) First aid records (not including medical histories) of one-time treatment and subsequent observation of minor scratches, cuts, burns, splinters, and the like which do not involve medical treatment, loss of consciousness, restriction of work or motion, or transfer to another job, if made on-site by a non-physician and if maintained separately from the employer's medical program and its records, and

1910.1020(d)(1)(i)(C)

(C) The medical records of employees who have worked for less than one (1) year for the employer need not be retained beyond the term of employment if they are provided to the employee upon the termination of employment.

1910.1020(d)(1)(ii)

(ii) "Employee exposure records." Each employee exposure record shall be preserved and maintained for at least thirty (30) years, except that:

*[61 FR 5507, Feb. 13, 1996, 61 FR 9227, March 7, 1996, 61 FR 31427, June 20, 1996.]

1910.1020(d)(1)(ii)(A)

(A) Background data to environmental (workplace) monitoring or measuring, such as laboratory reports and worksheets, need only be retained for one (1) year so long as the sampling results, the collection methodology (sampling plan), a description of the analytical and mathematical methods used, and a summary of other background data relevant to interpretation of the results obtained, are retained for at least thirty (30) years, and

1910.1020(d)(1)(ii)(B)

(B) Material safety data sheets and paragraph (c)(5)(iv) records concerning the identity of a substance or agent need not be retained for any specified period as long as some record of the identity (chemical name if known) of the substance or agent, where it was used, and when it was used is retained for at least thirty (30) years(1).

Footnote(1) Material safety data sheets must be kept for those chemicals currently in use that are effected by the Hazard and Communication Standard in accordance with 29 CFR 1910.1200(g).

1910.1020(d)(1)(ii)(C)

(C) Biological monitoring results designated as exposure records by specific occupational safety and health standards shall be preserved and maintained as required by the specific standard.

1910.1020(d)(1)(iii)

(iii) "Analyses using exposure or medical records." Each analysis using exposure or medical records shall be preserved and maintained for at least thirty (30) years.

1910.1020(d)(2)

(2) Nothing in this section is intended to mandate the form, manner, or process by which an employer preserves a record so long as the information contained in the record is preserved and retrievable, except that chest X-ray films shall be preserved in their original state.

1910.1020(e)(1)(i)

(e) "Access to records" (1) "General." (i) Whenever an employee or designated representative requests access to a record, the employer shall assure that access is provided in a reasonable time, place, and manner. If the employer cannot reasonably provide access to the record within fifteen (15) working days, the employer shall within the fifteen (15) working days apprise the employee or designated representative requesting the record of the reason for the delay and the earliest date when the record can be made available.

1910.1020(e)(1)(ii)

(ii) The employer may require of the requester only such information as should be readily known to the requester and which may be necessary to locate or identify the records being requested (e.g., dates and locations where the employee worked during the time period in question).

1910.1020(e)(1)(iii)

(iii) Whenever an employee or designated representative requests a copy of a record, the employer shall assure that either:

1910.1020(e)(1)(iii)(A)

(A) A copy of the record is provided without cost to the employee or representative,

1910.1020(e)(1)(iii)(B)

(B) The necessary mechanical copying facilities (e.g., photocopying) are made available without cost to the employee or representative for copying the record, or

1910.1020(e)(1)(iii)(C)

(C) The record is loaned to the employee or representative for a reasonable time to enable a copy to be made.

1910.1020(e)(1)(iv)

(iv) In the case of an original X-ray, the employer may restrict access to on-site examination or make other suitable arrangements for the temporary loan of the X-ray.

1910.1020(e)(1)(v)

(v) Whenever a record has been previously provided without cost to an employee or designated representative, the employer may charge reasonable, non-discriminatory administrative costs (i.e., search and copying expenses but not including overhead expenses) for a request by the employee or designated representative for additional copies of the record, except that

1910.1020(e)(1)(v)(A)

(A) An employer shall not charge for an initial request for a copy of new information that has been added to a record which was previously provided; and

1910.1020(e)(1)(v)(B)

(B) An employer shall not charge for an initial request by a recognized or certified collective bargaining agent for a copy of an employee exposure record or an analysis using exposure or medical records.

1910.1020(e)(1)(vi)

(vi) Nothing in this section is intended to preclude employees and collective bargaining agents from collectively bargaining to obtain access to information in addition to that available under this section.

1910.1020(e)(2)(i)(A)

(2) "Employee and designated representative access" (i) "Employee exposure records." (A) Except as limited by paragraph (f) of this section, each employer shall, upon request, assure the access to each employee and designated representative to employee exposure records relevant to the employee. For the purpose of this section, an exposure record relevant to the employee consists of:

1910.1020(e)(2)(i)(A){1}

{1} A record which measures or monitors the amount of a toxic substance or harmful physical agent to which the employee is or has been exposed;

1910.1020(e)(2)(i)(A){2}

{2} In the absence of such directly relevant records, such records of other employees with past or present job duties or working conditions related to or similar to those of the employee to the extent necessary to reasonably indicate the amount and nature of the toxic substances or harmful physical agents to which the employee is or has been subjected, and

1910.1020(e)(2)(i)(A){3}

{3} Exposure records to the extent necessary to reasonably indicate the amount and nature of the toxic substances or harmful physical agents at workplaces or under working conditions to which the employee is being assigned or transferred.

1910.1020(e)(2)(i)(B)

(B) Requests by designated representatives for unconsented access to employee exposure records shall be in writing and shall specify with reasonable particularity:

1910.1020(e)(2)(i)(B){1}

{1} The record requested to be disclosed; and

1910.1020(e)(2)(i)(B){2}

{2} The occupational health need for gaining access to these records.

1910.1020(e)(2)(ii)(A)

(ii) "Employee medical records." (A) Each employer shall, upon request, assure the access of each employee to employee medical records of which the employee is the subject, except as provided in paragraph (e)(2)(ii)(D) of this section.

1910.1020(e)(2)(ii)(B)

(B) Each employer shall, upon request, assure the access of each designated representative to the employee medical records of any employee who has given the designated representative specific written consent. Appendix A to this section contains a sample form which may be used to establish specific written consent for access to employee medical records.

1910.1020(e)(2)(ii)(C)

(C) Whenever access to employee medical records is requested, a physician representing the employer may recommend that the employee or designated representative:

1910.1020(e)(2)(ii)(C){1}

{1} Consult with the physician for the purposes of reviewing and discussing the records requested,

1910.1020(e)(2)(ii)(C){2}

{2} Accept a summary of material facts and opinions in lieu of the records requested, or

1910.1020(e)(2)(ii)(C){3}

{3} Accept release of the requested records only to a physician or other designated representative.

1910.1020(e)(2)(ii)(D)

(D) Whenever an employee requests access to his or her employee medical records, and a physician representing the employer believes that direct employee access to information contained in the records regarding a specific diagnosis of a terminal illness or a psychiatric condition could be detrimental to the employee's health, the employer may inform the employee that access will only be provided to a designated representative of the employee having specific written consent, and deny the employee's request for direct access to this information only. Where a designated representative with specific written consent requests access to information so withheld, the employer shall assure the access of the designated representative to this information, even when it is known that the designated representative will give the information to the employee.

1910.1020(e)(2)(ii)(E)

(E) A physician, nurse, or other responsible health care personnel maintaining employee medical records may delete from requested medical records the identity of a family member, personal friend, or fellow employee who has provided confidential information concerning an employee's health status.

1910.1020(e)(2)(iii)(A)

(iii) Analyses using exposure or medical records. (A) Each employer shall, upon request, assure the access of each employee and designated representative to each analysis using exposure or medical records concerning the employee's working conditions or workplace.

1910.1020(e)(2)(iii)(B)

(B) Whenever access is requested to an analysis which reports the contents of employee medical records by either direct identifier (name, address, social security number, payroll number, etc.) or by information which could reasonably be used under the circumstances indirectly to identify specific employees (exact age, height, weight, race, sex, date of initial employment, job title, etc.), the employer shall assure that personal identifiers are removed before access is provided. If the employer can demonstrate that removal of personal identifiers from an analysis is not feasible, access to the personally identifiable portions of the analysis need not be provided.

1910.1020(e)(3)(i)

(3) "OSHA access." (i) Each employer shall, upon request, and without derogation of any rights under the Constitution or the Occupational Safety and Health Act of 1970, 29 U.S.C. 651 "et seq.," that the employer chooses to exercise, assure the prompt access of representatives of the Assistant Secretary of Labor for Occupational Safety and Health to employee exposure and medical records and to analyses using exposure or medical records. Rules of agency practice and procedure governing OSHA access to employee medical records are contained in 29 CFR 1913.10.

1910.1020(e)(3)(ii)

(ii) Whenever OSHA seeks access to personally identifiable employee medical information by presenting to the employer a written access order pursuant to 29 CFR 1913.10(d), the employer shall prominently post a copy of the written access order and its accompanying cover letter for at least fifteen (15) working days.

1910.1020(f)(1)

(f) "Trade secrets." (1) Except as provided in paragraph (f)(2) of this section, nothing in this section precludes an employer from deleting from records requested by a health professional, employee, or designated representative any trade secret data which discloses manufacturing processes, or discloses the percentage of a chemical substance in mixture, as long as the health professional, employee, or designated representative is notified that information has been deleted. Whenever deletion of trade secret information substantially impairs evaluation of the place where or the time when exposure to a toxic substance or harmful physical agent occurred, the employer shall provide alternative information which is sufficient to permit the requesting party to identify where and when exposure occurred.

1910.1020(f)(2)

(2) The employer may withhold the specific chemical identity, including the chemical name and other specific identification of a toxic substance from a disclosable record provided that:

1910.1020(f)(2)(i)

(i) The claim that the information withheld is a trade secret can be supported;

1910.1020(f)(2)(ii)

(ii) All other available information on the properties and effects of the toxic substance is disclosed;

1910.1020(f)(2)(iii)

(iii) The employer informs the requesting party that the specific chemical identity is being withheld as a trade secret; and

1910.1020(f)(2)(iv)

(iv) The specific chemical identity is made available to health professionals, employees and designated representatives in accordance with the specific applicable provisions of this paragraph.

1910.1020(f)(3)

(3) Where a treating physician or nurse determines that a medical emergency exists and the specific chemical identity of a toxic substance is necessary for emergency or first-aid treatment, the employer shall immediately disclose the specific chemical identity of a trade secret chemical to the treating physician or nurse, regardless of the existence of a written statement of need or a confidentiality agreement. The employer may require a written statement of need and confidentiality agreement, in accordance with the provisions of paragraphs (f)(4) and (f)(5), as soon as circumstances permit.

1910.1020(f)(4)

(4) In non-emergency situations, an employer shall, upon request, disclose a specific chemical identity, otherwise permitted to be withheld under paragraph (f)(2) of this section, to a health professional, employee, or designated representative if:

1910.1020(f)(4)(i)

(i) The request is in writing;

1910.1020(f)(4)(ii)

(ii) The request describes with reasonable detail one or more of the following occupational health needs for the information:

1910.1020(f)(4)(ii)(A)

(A) To assess the hazards of the chemicals to which employees will be exposed;

1910.1020(f)(4)(ii)(B)

(B) To conduct or assess sampling of the workplace atmosphere to determine employee exposure levels;

1910.1020(f)(4)(ii)(C)

(C) To conduct pre-assignment or periodic medical surveillance of exposed employees;

1910.1020(f)(4)(ii)(D)

(D) To provide medical treatment to exposed employees;

1910.1020(f)(4)(ii)(E)

(E) To select or assess appropriate personal protective equipment for exposed employees;

1910.1020(f)(4)(ii)(F)

(F) To design or assess engineering controls or other protective measures for exposed employees; and

1910.1020(f)(4)(ii)(G)

(G) To conduct studies to determine the health effects of exposure.

1910.1020(f)(4)(iii)

(iii) The request explains in detail why the disclosure of the specific chemical identity is essential and that, in lieu thereof, the disclosure of the following information would not enable the health professional, employee or designated representative to provide the occupational health services described in paragraph (f)(4)(ii) of this section;

1910.1020(f)(4)(iii)(A)

(A) The properties and effects of the chemical;

1910.1020(f)(4)(iii)(B)

(B) Measures for controlling workers' exposure to the chemical;

1910.1020(f)(4)(iii)(C)

(C) Methods of monitoring and analyzing worker exposure to the chemical; and

1910.1020(f)(4)(iii)(D)

(D) Methods of diagnosing and treating harmful exposures to the chemical;

1910.1020(f)(4)(iv)

(iv) The request includes a description of the procedures to be used to maintain the confidentiality of the disclosed information; and

1910.1020(f)(4)(v)

(v) The health professional, employee, or designated representative and the employer or contractor of the services of the health professional or designated representative agree in a written confidentiality agreement that the health professional, employee or designated representative will not use the trade secret information for any purpose other than the health need(s) asserted and agrees not to release the information under any circumstances other than to OSHA, as provided in paragraph (f)(9) of this section, except as authorized by the terms of the agreement or by the employer.

1910.1020(f)(5)

(5) The confidentiality agreement authorized by paragraph (f)(4)(iv) of this section:

1910.1020(f)(5)(i)

(i) May restrict the use of the information to the health purposes indicated in the written statement of need;

1910.1020(f)(5)(ii)

(ii) May provide for appropriate legal remedies in the event of a breach of the agreement, including stipulation of a reasonable pre-estimate of likely damages; and,

1910.1020(f)(5)(iii)

(iii) May not include requirements for the posting of a penalty bond.

1910.1020(f)(6)

(6) Nothing in this section is meant to preclude the parties from pursuing non-contractual remedies to the extent permitted by law.

1910.1020(f)(7)

(7) If the health professional, employee or designated representative receiving the trade secret information decides that there is a need to disclose it to OSHA, the employer who provided the information shall be informed by the health professional prior to, or at the same time as, such disclosure.

1910.1020(f)(8)

(8) If the employer denies a written request for disclosure of a specific chemical identity, the denial must:

1910.1020(f)(8)(i)

(i) Be provided to the health professional, employee or designated representative within thirty days of the request;

1910.1020(f)(8)(ii)

(ii) Be in writing;

1910.1020(f)(8)(iii)

(iii) Include evidence to support the claim that the specific chemical identity is a trade secret;

1910.1020(f)(8)(iv)

(iv) State the specific reasons why the request is being denied; and,

1910.1020(f)(8)(v)

(v) Explain in detail how alternative information may satisfy the specific medical or occupational health need without revealing the specific chemical identity.

1910.1020(f)(9)

(9) The health professional, employee, or designated representative whose request for information is denied under paragraph (f)(4) of this section may refer the request and the written denial of the request to OSHA for consideration.

1910.1020(f)(10)

(10) When a health professional, employee, or designated representative refers a denial to OSHA under paragraph (f)(9) of this section, OSHA shall consider the evidence to determine if:

1910.1020(f)(10)(i)

(i) The employer has supported the claim that the specific chemical identity is a trade secret;

1910.1020(f)(10)(ii)

(ii) The health professional employee, or designated representative has supported the claim that there is a medical or occupational health need for the information; and

1910.1020(f)(10)(iii)

(iii) The health professional, employee or designated representative has demonstrated adequate means to protect the confidentiality.

1910.1020(f)(11)(i)

(11)(i) If OSHA determines that the specific chemical identity requested under paragraph (f)(4) of this section is not a "bona fide" trade secret, or that it is a trade secret but the requesting health professional, employee or designated representatives has a legitimate medical or occupational health need for the information, has executed a written confidentiality agreement, and has shown adequate means for complying with the terms of such agreement, the employer will be subject to citation by OSHA.

1910.1020(f)(11)(ii)

(ii) If an employer demonstrates to OSHA that the execution of a confidentiality agreement would not provide sufficient protection against the potential harm from the unauthorized disclosure of a trade secret specific chemical identity, the Assistant Secretary may issue such orders or impose such additional limitations or conditions upon the disclosure of the requested chemical information as may be appropriate to assure that the occupational health needs are met without an undue risk of harm to the employer.

1910.1020(f)(12)

(12) Notwithstanding the existence of a trade secret claim, an employer shall, upon request, disclose to the Assistant Secretary any information that this section requires the employer to make available. Where there is a trade secret claim, such claim shall be made no later than at the time the information is provided to the Assistant Secretary so that suitable determinations of trade secret status can be made and the necessary protections can be implemented.

1910.1020(f)(13)

(13) Nothing in this paragraph shall be construed as requiring the disclosure under any circumstances of process or percentage of mixture information that is a trade secret.

1910.1020(g)(1)

(g) "Employee information." (1) Upon an employee's first entering into employment, and at least annually thereafter, each employer shall inform current employees covered by this section of the following:

1910.1020(g)(1)(i)

(i) The existence, location, and availability of any records covered by this section;

1910.1020(g)(1)(ii)

(ii) The person responsible for maintaining and providing access to records; and

1910.1020(g)(1)(iii)

(iii) Each employee's rights of access to these records.

1910.1020(g)(2)

(2) Each employer shall keep a copy of this section and its appendices, and make copies readily available, upon request, to employees. The employer shall also distribute to current employees any informational materials concerning this section which are made available to the employer by the Assistant Secretary of Labor for Occupational Safety and Health.

1910.1020(h)(1)

(h) "Transfer of records." (1) Whenever an employer is ceasing to do business, the employer shall transfer all records subject to this section to the successor employer. The successor employer shall receive and maintain these records.

1910.1020(h)(2)

(2) Whenever an employer is ceasing to do business and there is no successor employer to receive and maintain the records subject to this standard, the employer shall notify affected current employees of their rights of access to records at least three (3) months prior to the cessation of the employer's business.

1910.1020(h)(3)

(3) Whenever an employer either is ceasing to do business and there is no successor employer to receive and maintain the records, or intends to dispose of any records required to be preserved for at least thirty (30) years, the employer shall:

1910.1020(h)(3)(i)

(i) Transfer the records to the Director of the National Institute for Occupational Safety and Health (NIOSH) if so required by a specific occupational safety and health standard; or

1910.1020(h)(3)(ii)

(ii) Notify the Director of NIOSH in writing of the impending disposal of records at least three (3) months prior to the disposal of the records.

1910.1020(h)(4)

(4) Where an employer regularly disposes of records required to be preserved for at least thirty (30) years, the employer may, with at least (3) months notice, notify the Director of NIOSH on an annual basis of the records intended to be disposed of in the coming year.

1910.1020(i)

(i) "Appendices." The information contained in appendices A and B to this section is not intended, by itself, to create any additional obligations not otherwise imposed by this section nor detract from any existing obligation.

IV. BLOODBORNE PATHOGENS

A. INTRODUCTION

The Bloodborne Pathogens Standard was the first official step OSHA took in usurping the health-care industry within its active domain. More than anything, it helped to mandate practices that were generally accepted but not uniformly implemented, giving the force of regulation to Centers for Disease Control (CDC)-recommended guidelines. In November of 1999, OSHA released a Bloodborne Pathogens Compliance Directive at the bequest of strong labor urging OSHA to ensure compliance with certain aspects of the rule, which they felt employers were not prioritizing or addressing adequately, mainly, needle stick prevention. Thus, inspections going forward will focus even more specifically on the formal and complete annual review by management of alternative safer sharps and sharps-containment equipment and technologies, consideration of needleless systems, disinfection schedules, medical waste handling, etc.

B. DISCUSSION

The author finds that clients have usually implemented the requirements of the Bloodborne Pathogens Standard better than any other OSHA requirements, obviously because of the relevance to the operation, and the qualified personnel available. Because of this, no comprehensive Bloodborne Pathogens Guide has been developed until this manual. There are a handful of pitfalls that have been noticed in the Exposure Control Plan in some facilities, which are as follows:

- Ensuring complete and adequate bloodborne training is provided. Some training videos have missing components and employees must have an opportunity to ask questions.
- Actually providing the hepatitis B virus (HBV) vaccine and tracking people who have asked for it for their subsequent booster shots or into maintaining signed declination forms on file.
- Assuring that postexposure medical records are available on site.
- Ensuring that "exposure incidents" are *accurately identified*, documented, and tracked as such on the OSHA 200 Log (illness side).
- Ensuring that the medical follow-up and counseling are completed and documented (including signatures of declination if staff refuses any recommended steps).
- Ensuring that these medical follow-up and counseling documents are available on-site.
- If overconservative actions similar to those taken for "exposure incidents" were taken for an incident that was not a true exposure incident, ensuring that these cases are distinguishable. Do not allow these cases to be mistaken for actual exposure incidents, which are to be recorded on the OSHA 200 Log, *require* the follow-up, and act as an inspection flag.

Example: If the administrative system treats incidents of urine splash to the eye (which is not an "exposure incident" unless there was visible blood involved), along similar lines of legitimate "exposure incidents," then the facility must be clear, *not to report these as such*, and the documentation must be clear, not to confuse and mislead the inspector into believing that the facility has underreported exposure incidents, or has reported additional numbers of exposure incidents. This has caused some confusion in a number of facilities and has complicated inspections unnecessarily.

With the advent of the Bloodborne Compliance Directive, the following items should also ensure that the directive is implemented and *well documented:*

- Annual review of new and existing medical equipment and technology that may decrease the risk of bloodborne exposure to staff. (Employers need to have medical documentation

prescribing a particular device as medically beneficial to the patient if there is a safer device available that is not used.)
- Adequate number of sharps containers.
- Logistically appropriate placement of sharps containers (a common cause of needle-stick injuries is requiring the health care worker to walk distances to dispose of the sharp).
- Proper design of containers given the work and conditions.
- The prompt removal of filled containers and ensuring there is no "overfilling" (which is a major cause of needle sticks among health care workers).
- Adequate disinfection schedule and procedures.
- Adequate disinfectant and disinfect procedures.
- Proper and adequate medical waste handling.

C. SUMMARY

Ensure that the Exposure Control Plan (a sample plan is provided in Section IV.D) and all its elements are updated and in place, in addition to the above, include the following:

- The practice of universal/standard precautions
- The availability and proper use of personal protective equipment
- Initial and annual staff training
- Accessible and available adequate spill kits
- Adequate disinfecting regime
- Proper use of biohazard signs and labels
- Proper use and handling of sharps and sharps containers
- Proper use and handling of medical wastes and contaminated materials

Section IV provides a comprehensive bloodborne pathogens compliance guide for long-term care (based on a variety of materials, primarily, the Employer Guide and Model developed by the New York State Department of Labor).

D. SAMPLE PLAN: THE EXPOSURE CONTROL PLAN

The _____ [facility] is committed to providing a safe and healthy environment for our staff, residents, and visitors. This Exposure Control Plan (ECP) is designed to eliminate or reduce the risk of occupational exposure to and transmission of bloodborne pathogens, as much as possible. This ECP specifically covers all the aspects of OSHA Bloodborne Pathogens Standard, 29 CFR 1910.130. It operates in conjunction with the facility's overall Infection Control Plan, which also focuses on general infection control and prevention and, of course, resident transmission.

Those employees who are reasonably anticipated to have contact with, or exposure to, blood or other potentially infected materials are required to comply with the procedures and work practices outlined herein.

This ECP serves as an essential document and implementation tool to ensure compliance and includes the following elements:

1. Employee Exposure Determination
2. Definitions to Determine Exposure
3. Implementing Methods of Control
   a. Universal/Standard Precautions
   b. Engineering Controls and Work Practices
   c. Personal Protective Equipment
   d. Hepatitis B Virus Vaccine

      e. Postexposure Evaluation Follow-Up
      f. Housekeeping and Laundry
      g. Labeling
      h. Training and Communicating Hazards to Employees
   4. Record Keeping

Sample Documents

Program Administrator(s)

_____ [title] is/are responsible for the implementation of the ECP. _____ [title] will maintain and update the written ECP at least annually and whenever necessary to include new or modified tasks and procedures which affect occupational exposure and to reflect new or revised employee positions with occupational exposure. _____ [title] will have the responsibility for written housekeeping protocols and will ensure that effective disinfectants are purchased. _____ [title] will be responsible for ensuring that all medical actions required are performed and that appropriate medical records are maintained. _____ [title] will be responsible for training, documentation of training, and making the written ECP available to employees, OSHA and other appropriate representatives. _____ [title] will maintain and provide all necessary personal protective equipment (PPE), engineering controls (sharps containers, etc.), labels, and red bags as required by this standard. _____ [title] will ensure that adequate supplies of the preceding equipment are readily available.

[*Note:* A single Program Administrator can be identified to simplify matters, particularly if the facility is a smaller operation and the program responsibilities are held by one individual, then combine the above items.]

1. Employee Exposure Determination

[*Note:* In making exposure determinations, the assessment must be made without regard to the use of personal protective equipment. In other words, assess job tasks risks as if they were executed without the benefit of gloves, or goggles, etc.]

Occupations that may involve risk from occupational exposure to blood or other potentially infectious materials are physician; physicians assistant; nurse; phlebotomist; medical examiner; emergency medical technician (EMT), supervisor (performing first aid); dentist; dental hygienist; medical technician; regulated medical waste handler; some laundry and housekeeping employees; life guards; etc.

2. Definitions to Determine Exposure

- *Blood*—Human blood, human blood components, and products made from human blood.
- *Bloodborne Pathogens*—Pathogenic microorganisms that are present in human blood and can infect and cause disease in humans. These pathogens include, but are not limited to, hepatitis B virus (HBV), hepatitis C virus (HCV), human immunodeficiency virus (HIV), syphilis, etc.
- *Contaminated*—The presence *or the reasonably anticipated presence* of blood or other potentially infectious materials on an item or surface.
- *Exposure Incident*—A specific eye, mouth, other mucous membrane, nonintact skin, or parenteral contact with blood or other potentially infectious materials *that results* from the performance of an employee's duties.
- *Occupational Exposure*—*Reasonably anticipated* skin, eye, mucous membrane, or parenteral contact with blood or other potentially infectious materials that may result from the performance of an employee's duties.
- *Other Potentially Infectious Materials (OPIM)*—The following human body fluids: semen, vaginal secretions, cerebrospinal fluid, synovial fluid, pleural fluid, pericardial

fluid, peritoneal fluid, amniotic fluid, saliva in dental procedures, *any body fluid visibly contaminated with blood,* all body fluids in situations where it is difficult or impossible to differentiate between body fluids, any unfixed tissue or organ (other than intact skin) from a human (living or dead), HIV-containing cells or tissue cultures, organ cultures, and HIV- or HBV- or HCV-containing culture medium or other solutions, and blood, organs, or other tissue from experimental animals infected with HIV or HBV/HCV.

* *Regulated Waste*
  —Liquid or semiliquid blood or other potentially infectious materials,
  —Contaminated items that would release blood or other potentially infectious materials in liquid or semiliquid state if compressed,
  —Items that are caked with dried blood or other potentially infectious materials and are capable of releasing these materials during handling,
  —Contaminated sharps;
  —Pathological/microbiological wastes with blood or other potentially infectious materials.

[*Note:* Blood-stained material, in and of itself, may not be deemed a regulated waste if it would not release blood or OPIM in a liquid or semiliquid state if compressed, or if it is not capable of releasing these materials during handling. Therefore, stained bandages, laundry, clothing, sanitary napkins, etc. of that nature would not have to be handled as regulated medical waste, and is ordinary municipal waste.]

Make determinations on the occupational risks for each job category and/or tasks within the context of the above definitions. Again, make these determinations as if exposure were to occur without the use of personal protective equipment. Complete both, or only one, of the following sections as they apply to your operation.

List all job classifications in which employees have occupational exposure:

Job Title	Department/Location

[continue on separate sheet]

List job classifications in which *some* employees have occupational exposure, including a list of tasks and procedures in which occupational exposure may occur for these individuals (i.e., custodians occasionally cleaning contaminated equipment, laundries where some workers are assigned the task of handling contaminated laundry, etc.):

Job Title	Department/Location	Task Procedure

[continue on separate sheet]

Contractors employees
The facility will notify contract employers (plumber, etc.) of potential contact with blood or other potentially infectious materials so they can take appropriate precautions.

"Good Samaritan" acts
Good Samaritan acts resulting in exposure to blood or other potentially infectious materials from assisting a fellow employee (i.e., assisting a co-worker with a nosebleed, giving CPR, or first aid), when it is not part of the job description, a primary, or secondary duty, are not included in the Bloodborne Pathogens Standard. It is encouraged that employers offer postexposure evaluation and follow-up on such cases, nonetheless.

3. Implementing Methods of Control

a. Universal/standard precautions
All employees must utilize Universal or Standard Precautions (revised name for this protocol), an infection-control method that requires employees to *assume that all human blood and specified human body fluids* are infectious for HIV, HBV/HCV, and other bloodborne pathogens, and *must be treated accordingly.*

b. Engineering controls and work practices
Engineering controls and work practice controls will be used to prevent or minimize exposure to bloodborne pathogens (i.e., sharps containers, self-sheathing needles, puncture-resistant disposal containers for contaminated sharps, orthodontia wire, or broken glass, mechanical needle recapping devices, biosafety cabinets, ventilated laboratory hoods, etc.), along with the physical logistics and planning of their use.

Examples of work practice controls include, but are not limited to:

- Providing readily accessible hand-washing facilities
- Washing hands immediately or as soon as feasible after removal of gloves
- At nonfixed sites that lack hand-washing facilities, provide interim hand-washing measures, such as antiseptic, towelettes, and paper towels. Employees can later wash their hands with soap and water as soon as feasible
- Washing body parts as soon as possible after skin contact with blood or OPIM occurs
- Prohibiting the recapping or bending of needles
- Shearing or breaking contaminated needles is prohibited
- Proper use of labeling
- Equipment decontamination
- Prohibiting eating, drinking, smoking, applying cosmetics or lip balm, and handling contact lenses in work areas where there is a likelihood of occupational exposure
- Prohibiting food/drink from being kept in refrigerators, freezers, shelves, cabinets, or on countertops or benchtops where blood or other potentially infectious materials are present
- Requiring that all procedures involving blood or OPIM shall be performed in such a manner to minimize splashing, splattering, and generation of droplets of these substances
- Placing specimens of blood or OPIM in a container, which prevents leakage during collection, handling, processing, storage, transport, or shipping
- Examining equipment that may become contaminated with blood or OPIM prior to servicing or shipping and decontaminating such equipment as necessary. Items will be labeled per the standard if not completely decontaminated.

The specific *engineering controls* and *work practice controls* used in this facility, and where they will be used are as listed below:

Engineering Controls/Work Practice Controls	Location

[continue on separate sheet]

New technology for needles and sharps will be evaluated and implemented whenever necessary and at least annually to prevent further accidental needle sticks and cuts. The facility's engineering controls will be inspected and maintained or replaced by [title] _____ every _____.
Note: Define a schedule and specify a person responsible for examining the effectiveness of the engineering controls used. A time period must also be stated for the inspection of sharps containers, to *ensure that the containers are not overloaded.* It is recommended that a margin of safety be incorporated when determining this inspection interval.

c. *Personal protective equipment (PPE)*

PPE must be used if occupational exposure remains after instituting engineering and work practice controls, or if controls are not feasible. The type and characteristics of the PPE will depend upon the task and degree of exposure anticipated. PPE items include gloves, gowns, lab coats, face shields, masks, eye protection (splash-proof goggles, safety glasses with side shields), resuscitation bags and mouthpiece, etc.

Training will be provided by _____ in the use of appropriate PPE for employees' specific job classifications and task/procedures they will perform. Additional training will be provided, whenever necessary, such as, if employees take a new position, or if new duties are added to their current position.

First aid responders should have quick access to kits with impervious gloves, resuscitation bags or mouthpieces, eye protection, aprons, disinfectant towelettes for hand washing, and red bags or biohazard-labeled bags.

Appropriate PPE is required for the following tasks with this specific equipment:

Task	Equipment

[continue on separate sheet]

[Specify what equipment will be issued, how, when and who will provide the PPE.]

All employees using PPE must observe the following precautions:

- Wash hands immediately or as soon as feasible after removal of gloves or other PPE.
- Remove PPE before leaving the work area and after a garment becomes contaminated.
- Place used PPE in appropriately designated area or container when being stored, washed, decontaminated, or discarded.

Designated Areas and/or Containers	Location

_____   _____
_____   _____
_____   _____
_____   _____

[continue as necessary]

- Wear appropriate gloves when it can be reasonably anticipated that you may have contact with blood or OPIM and when handling or touching contaminated items or surfaces. Replace gloves if torn, punctured, contaminated, or if their ability to function as a barrier is compromised.
- Following any contact of body areas with blood or any other infectious materials, you must wash your hands and any other exposed skin with soap and water as soon as possible. Employees must also flush exposed mucous membranes (eyes, mouth, etc.) with copious amounts of water.
- Utility gloves may be decontaminated for reuse if their integrity is not compromised. The decontamination procedure will consist of _____. Discard utility gloves when they show signs of cracking, peeling, tearing, puncturing, or deterioration.
- Never wash or decontaminate disposable gloves for reuse or before disposal.
- Wear appropriate face and eye protection such as a mask with glasses with solid side shields or a chin-length face shield when splashes, sprays, splatters, or droplets of blood or OPIM pose a hazard to the eye, nose, or mouth.
- If a garment is penetrated by blood and other potentially infectious materials, the garment(s) must be removed immediately or as soon as feasible. If a pullover scrub (as opposed to scrubs with snap closures) becomes minimally contaminated, employees should be trained to remove the pullover scrub in such a way as to avoid contact with the outer surface (i.e., rolling up the garment as it is pulled toward the head for removal); however, if the amount of blood exposure is such that the blood penetrates the scrub and contaminates the inner surface, not only is it impossible to remove the scrub without exposure to blood, the penetration itself would constitute an exposure. It may be prudent to train employees to cut such a contaminated scrub to aid removal and prevent exposure to the face.
- Repair and/or replacement of PPE will be at no cost to employees.

d. Hepatitis B virus (HBV) vaccine

_____ [title] will provide information on hepatitis B vaccinations addressing their safety, benefits, efficacy, methods of administration, and availability. The HBV vaccine series will be made available at no cost within 10 days of initial assignment to employees who have occupational exposure to blood or OPIM, unless:

- The employee has previously received the series.
- Antibody testing reveals that the employee is immune.
- Medical reasons present taking the vaccine.
- The employee chooses not to participate.

All employees are strongly encouraged to receive the HBV vaccine series. However, if an employee chooses to decline HBV vaccination, then the employee must sign a statement to this effect. The declination form must include the wording mandated by the Standard to state:

I understand that due to my occupational exposure to blood or other potentially infectious material, I may be at risk of acquiring Hepatitis B virus (HBV) infection. I have been given the opportunity to be vaccinated with Hepatitis B vaccine, at no charge to myself. However, I decline Hepatitis B vaccination at this time. I understand that by declining this vaccine, I continue to be at risk of acquiring Hepatitis B, a serious disease. If in the future I continue to have occupational exposure to blood or other potentially infectious materials and I want to be vaccinated with Hepatitis B vaccine, I can receive the vaccination series at no charge to me. [employee signature and date]

Employees who decline may request and obtain the vaccination at a later date at no cost. Documentation or refusal of the HBV vaccination will be kept in _____ with the employee's other medical records.

Participation in prescreening is not a prerequisite for receiving the HBV vaccine. Vaccination must be administered in accordance with U.S. Public Health Service (USPHS)-recommended protocols, and booster doses must be available to employees as recommended by USPHS. See sample "Hepatitis B Vaccine Immunization Record" form [at end of Section IV.D]

e.   Postexposure evaluation follow-up

Should any exposure incident occur, contact _____ [title] immediately. Each exposure must be documented by the employee on an "Exposure Report Form" [sample form at end of this section] _____ [title] will add any additional information as needed.

An immediately available confidential medical evaluation and follow-up will be conducted by _____. The following elements will be performed:

- Document the routes of exposure and how exposure occurred.
- Identify and document the source individual, unless the employer can establish that identification is infeasible or prohibited by state or local law.
- Obtain consent and test source individual's blood as soon as possible to determine HIV and HBV/HCV infectivity and document the source's blood test results.
- If the source individual is known to be infected with either HIV or HBV/HCV, testing need not be repeated to determine the known infectivity.
- Provide the exposed employee with the source individual's test results and information about applicable disclosure laws and regulations concerning the source identity and infectious status.
- After obtaining consent, collect exposed employee's blood as soon as feasible after the exposure incident and test blood for HBV/HCV and HIV serological status.
- If the employee does not give consent for HIV serological testing during the collection of blood for baseline testing, preserve the baseline blood sample for at least 90 days.
- Refer employee to appropriate health care personnel to determine if postexposure prophylaxis (PEP) is recommended per CDC guidelines.
- Assure that the employee starts PEP if it is recommended and he or she consents to the treatment.
- Assure that the employee receives medical counseling on the evaluation results, and any follow-up needed, including risk of infection, symptoms of HIV/HBV/HCV disease, continued treatment, prophylaxis, and reevaluation phases.
- Provide the Exposure Incident Report, the Request for Source Individual Evaluation, and the Employee Exposure Follow-up Record (samples attached) to the employee so he or she may bring them with any additional relevant medical information to the medical evaluation.
- Maintain original copies of these with employee's medical records.

Timeliness: Prompt medical evaluation and prophylaxis is imperative, following an exposure incident. Timeliness is, therefore, an important factor in effective medical treatment. In the May 15, 1998 CDC issue of MWWR Recommendations and Reports, the CDC recommends that PEP, when applicable, should be initiated as soon as possible, *within a few hours rather than days.* In most instances, PEP is not recommended within the context of exposure incidents in long-term care because of the low risks involved. However, should an exposure incident occur that calls for PEP treatment initiation, then it should be ensured that it can be provided within a few hours (2 h). Some facilities have a collaborative agreement with a local 24-hour, 7-days-a-week emergency care provider that it will be able to conduct the postexposure evaluation and initiate the PEP, if necessary, or at least deliver the initial dose of the PEP when another qualified health care provider has conducted the postexposure evaluation, and has indicated that PEP is in order.

[*Medical Note:* In deciding how a facility will manage this requirement, keep in mind that the types of exposure incidents in long-term care settings are usually of very little risk. They generally fall into the category of "no increased risk." This includes mucous membrane exposure to blood or fluid containing visible blood or other potentially infectious fluid or tissue and skin exposures involving a high titer of HIV prolonged contact—an extensive area or compromised skin integrity to blood or fluid containing visible blood or other potentially infectious fluid or tissue. Under these circumstances, the guidelines for chemoprophylaxis *after exposure to patients with HIV or positive risk factors for HIV*, is still only to *offer* a two or three drug antiretroviral regimen, the third of which poses possible toxicity risk of an additional drug, which may not be warranted. Thus, in a long-term care setting, it would be the exception, rather than the rule, that PEP would be offered. Oftentimes, the medical evaluation may strongly not recommend PEP treatment because of existing contraindications to the drug regimen, i.e., kidney or liver problems, age, etc. Other times, employees refuse or end the PEP treatment prematurely because of side effects.

Therefore, design a plan that would ensure immediate treatment or initiation of PEP, in the rare instance that it is necessary and appropriate. But weigh that likelihood (which must include the facility size, nature of operation, community profile for risk factors, past experience, etc.) with the benefits and burden to the staff and system in maintaining such drugs on site (which are expensive and do expire). For most facilities, a collateral care provider outside the facility generally serves their needs best by providing around-the-clock and weekend service for both medical evaluation and provision of immediate treatment if necessary. Ensure that this collaborator is indeed competent and ready to deliver this service without notice.]

[*Legal Note:* Public Health Law (Article 27-F) requires information about AIDS and HIV be kept confidential. This law requires that anyone receiving an HIV test *must* sign a consent form first. The law strictly limits disclosure of HIV-related information. When disclosure of HIV-related information is authorized by a signed release, the person who has been given the information *must* keep it confidential. Redisclosure may occur with another authorized signed release. The law only applies to people and facilities providing health or social services. If consent is not obtained, the employer must show that legally required consent could not be obtained. Where consent is not required by law, the source individual's blood, if available, should be tested and the results documented. If, during this time, the exposed employee elects to have the baseline sample tested, testing shall be done as soon as feasible.]

_____ [title] will review the circumstances of the exposure incident to determine if procedures, protocols, and/or training need to be revised.

Healthcare professionals

_____ [title] will ensure that health care professionals responsible for employee's hepatitis vaccination and postexposure evaluation and follow-up be given a copy of the OSHA Bloodborne Standard. _____ [title] will also ensure that the health care professional evaluating an employee after an exposure incident receives the following:

- A description of the employee's job duties relevant to the exposure incident
- Routes of exposure
- Circumstances of exposure
- If possible, results of the source individual's blood test; and relevant employee medical records, including vaccination status

Healthcare professional's written opinion

_____ will provide the employee with a copy of the evaluating health care professional's written opinion within 15 days after completion of the evaluation. For HBV vaccinations, the health care professional's written opinion will be limited to whether the employee requires or has received the HBV vaccination, or booster shot. The written opinion for postexposure evaluation and follow-up will be limited to whether or not the employee has been informed of the results of the medical evaluation and any medical conditions that may require further evaluation and treatment. All other diagnoses must remain confidential and not be included in the written report to our facility.

If the employer is also the health care professional, then it must ensure that the results of the employee's postexposure evaluation remain confidential from his or her co-workers.

[See the "Hepatitis B Vaccine Immunization Record," "The Exposure Incident Report," "Request for Source Individual Evaluation," "Documentation and Identification of Source Individual," "Employee Exposure Follow-Up Records" sample forms provided at the end of this section. These are not mandatory forms, however, a system to convey such information must be in place. Also see the checklist for Postexposure Follow-Up at the end of this section.]

f.　Housekeeping and laundry

_____ [title] has developed and implemented a written schedule for cleaning and decontaminating work surfaces as indicated by the Standard.

Cleaning Schedule

Area	Scheduled Cleaning (Day/Time)	Cleaners and Disinfectants Used	Specific Instructions

[Include a housekeeping schedule and method of decontamination, as well as location of cleanup and decontamination supplies. A list of approved sterilants can be obtained from the EPA at 800-447-6349.]

Cleaning Schedule

Area	Scheduled Cleaning (Day/Time)	Cleaners and Disinfectants Used	Specific Instructions

Examples of housekeeping procedural tasks

- Decontaminate work surfaces with an appropriate disinfectant after completion of procedures, immediately when overtly contaminated, after any spill of blood or OPIM, and at the end of the work shift when surfaces have become contaminated since the last cleaning.
- Remove and replace protective coverings such as plastic wrap and aluminum foil when contaminated.
- Inspect and decontaminate, on a regular basis, reusable receptacles such as bins, pails, and cans that have a likelihood for becoming contaminated. When contamination is visible, clean and decontaminate receptacles immediately, or as soon as is feasible.

- Always use mechanical means such as tongs, forceps, or a brush and a dustpan to pick up contaminated broken glassware; never pick up with hands even if gloves are worn.
- Store or process reusable sharps in a way that ensures safe handling.
- Place regulated waste in closable and labeled or color-coded containers. When storing, handling, transporting or shipping, place other regulated waste in containers that are constructed to prevent leakage.
- When discarding contaminated sharps, place them in containers that are closable, puncture-resistant, appropriately labeled or color-coded, and leak-proof on the sides and bottom.
- Ensure that sharps containers are easily accessible to personnel and located as close as feasible to the immediate area where sharps are used or can be reasonably anticipated to be found. Sharps containers also must be kept upright throughout use, replaced routinely, closed when moved, and not allowed to be overfilled.
- Never manually open, empty, or clean reusable contaminated sharps disposal containers.
- Discard all regulated waste according to federal, state, and local regulations (i.e., liquid blood or OPIM, items contaminated with blood, OIPM that would release these substances in a liquid or semiliquid state if compressed, items caked with dried blood or other potentially infectious materials and capable of releasing these materials during handling, contaminated sharps, and pathological and microbiological wastes containing blood or OPIM).

Laundry: The following contaminated articles will be laundered:

_____
_____
_____
_____
_____

Laundering will be performed by_____ at _____
The following requirements must be met, with respect to contaminated laundry:

- Handle contaminated laundry as little as possible and with a minimum of agitation.
- Use appropriate PPE when handling contaminated laundry.
- Place wet contaminated laundry in leak-proof, labeled or color-coded containers before transporting.
- Bag contaminated laundry at its location of use.
- Never sort or rinse contaminated laundry in areas of its use.
- Use red laundry bags or those marked with the biohazard symbol unless universal/standard precautions are in use at the facility and all employees recognize the bags as contaminated and have been trained in handling the bags. [Specify which labeling system, red bags or biohazard labeling, will be used.]
- All generators of laundry must have been determined if the receiving facility used universal/standard precautions. If not in use, then clearly mark laundry sent off site with orange biohazard labels or use red bags. Leak-proof bags must be used when necessary to prevent soak-through or leakage. [Specify which labeling system, red bags or biohazard labeling, will be used.]
- When handling and/or sorting contaminated laundry, utility gloves and other appropriate PPE (i.e., aprons, mask, eye protection) shall be worn.
- Laundries must have sharps containers readily accessible because of the incidence of needles and sharps being unintentionally mixed with laundry.
- Linen soiled with blood or body fluids should be placed and transported in bags that prevent leakage. If hot water is used, linen should be washed with detergent in water at least

140 to 160°F for 25 min. If low-temperature (less than 140°F) laundry cycles are used, chemicals suitable for low-temperature washing at proper use concentration should be used.

[See sample "Information of Regulated Medical Waste" questionnaire at the end of this section to assist in assessing contractors for medical waste transport, handling, and disposal.]

g. Labeling
The following labeling method(s) will be used at this facility:

_____
_____
_____
_____

_____ [title] will ensure warning labels are affixed or red bags are used as required. Employees are to notify _____ [title] if they discover unlabeled regulated waste containers.

Facilities must specify which warning methods are used, and communicate this information to all employees. The Standard requires that fluorescent orange or orange-red warning labels be attached to:

- Containers of regulated waste;
- Refrigerators and freezers containing blood and OPIM;
- Sharps disposal containers;
- Laundry bags and containers;
- Contaminated equipment for repair (portion contaminated);
- Other containers used to store, transport, or ship blood or OPIM.

Labels are not required when:

- Red bags or red containers are used;
- Containers of blood, blood components, or blood products are labeled with their contents and have been released for transfusion or other clinical use;
- Individual containers of blood or other potentially infectious materials are placed in a labeled container during storage, transport, shipment, or disposal. The warning label must be fluorescent orange or orange-red, contain the biohazard symbol, and the word BIOHAZARD in a contrasting color, and be attached to each object by string, wire, adhesive, or other method to prevent loss or unintentional removal of the label.

h. Training and communicating hazards to employees
Employees covered by the Bloodborne Pathogens Standard will receive an explanation of this ECP during their orientation training session. It will be reviewed in their annual refresher training thereafter. All staff will have an opportunity to review the plan at any time during their work shifts, contact _____ [title]. A copy of this ECP will be made available free of charge and within 15 days of the request.

All employees who have, or are reasonably anticipated to have occupational exposure to bloodborne pathogens will receive training conducted by _____ [title]. Training will be provided on the epidemiology of bloodborne pathogen diseases. Training will cover, at minimum, the following elements:

- A copy and explanation of the Standard;
- Epidemiology and symptoms of bloodborne pathogens;

- Modes of transmission;
- The ECP and how to obtain a copy;
- Methods to recognize exposure tasks/other activities that may involve exposure to blood;
- Use and limitations of engineering controls, work practices, and PPE;
- PPE—types, use, location, removal, handling, decontamination, and disposal;
- PPE—the basis for selection;
- Hepatitis B vaccine—offered free of charge (training will be given prior to vaccination on its safety, effectiveness, benefits, and method of administration);
- Emergency procedures for blood and OPIM;
- Exposure incident procedures;
- Postexposure evaluation and follow-up;
- Signs and labels and/or color coding;
- Question and answer session.

[*Note:* Make training materials, such as overheads, pictures, work sheets, pamphlets, etc. part of this ECP.]

An employee Education and Training Record will be completed for each employee upon completion of training. This document will be kept with the employee's records at_____. [See In-Service Records in Chapter 5.]

4. Record Keeping

Medical records

Medical records are maintained for each employee with occupational exposure in accordance with 29 CFR 1910.1020 (Employee Right to Access Medical and Exposure Records, previously 1910.20).

_____ [title] is responsible for maintenance of the required medical records and the records are kept at _____.

The medical record will include:

- The name and social security number of the employee,
- A copy of the employee's hepatitis B vaccinations and any medical records relative to the employee's ability to receive vaccination,
- A copy of all results of examinations, medical testing, and follow-up procedures as required by the Standard,
- A copy of all health care professional's written opinion(s) as required,

All employee medical records will be kept confidential and will not be disclosed or reported without the employee's express written consent to any person within or outside the workplace except as required by the Standard or as may be required by law. Employee medical records shall be maintained for at least the duration of employment plus 30 years in accordance to 29 CFR 1910.1020. Employee medical record shall be provided upon request of the employee or to anyone having written consent of the employee within 15 working days.

Transfer of records

If _____ [facility] ceases to do business and there is no successor employer to receive and retain the records for the prescribed period, the employer shall notify the Director of the National Institute for Occupational Safety and Health (NIOSH) at least 3 months prior to scheduled record disposal and prepare to transmit them to the Director.

Training records

Training records for Bloodborne Pathogens will be maintained by _____ at _____. The training record shall include:

- Dates of the training sessions;
- Contents or a summary of the training sessions;
- Names and qualifications of persons conducting the training;
- The names and job titles of all persons attending the training sessions.

Training records will be maintained for a minimum of 3 years from the date on which the training occurred. Employee training records will be provided upon request to the employee or the employee's authorized representative within 15 working days.

[Following are sample documents referred to in this sample ECP.]

SAMPLE FORM: HEPATITIS B VACCINE IMMUNIZATION RECORD

CONFIDENTIAL

HEPATITIS B VACCINE IMMUNIZATION RECORD

Vaccine is to be administered on: _____

Elected dates: _____

First: _____

One month from elected **date:** _____

Six months from elected **date:** _____

Employee name: _____

Date of first **dose:** _____

Date of second dose: _____

Date of third dose: _____

Antibody test results prevaccine (optional): _____

Antibody test results postvaccine (optional): _____

Time interval since last injection: _____

Employee signature: _____

SAMPLE FORM: EXPOSURE INCIDENT REPORT

ROUTES AND CIRCUMSTANCES OF EXPOSURE INCIDENT

PLEASE PRINT

Date completed: _____ Employee's name: _____

SS#: _____ Home Ph _____ Work Ph _____

DOB _____ Job title: _____

Employee vaccination status: _____

Date of exposure: _____ Time of exposure: _____ am _____ pm

Location of incident (home, street, clinic, etc.)—be specific: _____

Nature of incident (auto accident, trauma, medical emergency)—be specific:

_____

_____

_____

_____

Were you wearing personal protective equipment (PPE)? Yes____No____

If yes, list:

_____

Did the PPE fail? Yes _____ No _____

If yes, explain how:

_____

_____

_____

_____

What body fluid(s) were you exposed to (blood or other potentially infectious material)? Be specific:

_____

_____

What parts of your body became exposed? Be specific:_____

_____

Estimate the size of the area of your body that was exposed: _____

_____

For how long did the exposure last? _____

Did a foreign body (needle, nail, auto part, dental wire, broken glass, teeth, etc.) penetrate your body? Yes _____ No _____

If yes, what was the object? _____

Was any fluid injected into your body? Yes_____ No _____

If yes, what fluid? _____ How much? _____

Did you receive medical attention? Yes _____ No _____

If yes, where? _____

When?_____ By whom: _____

Identification of source individual(s) _____

Name(s) _____

Did you treat the patient directly? Yes _____ No _____

If yes, what treatment did you provide? Be specific: _____

_____

Other pertinent information: _____

Completed by: _____ Title: _____

SAMPLE LETTER: REQUEST FOR SOURCE INDIVIDUAL EVALUATION

Dear Physician [resident's assigned physician, or in-house medical director with authority to provide such information, or the source individual's health care provider if the person is a visitor, vendor, another employee, etc.]

[In the process of providing care to the patient/in an incident at our facility/a description of the general nature of the business at hand when the incident occurred,] one of our staff was involved in an event with your patient that may have resulted in exposure to a Bloodborne Pathogen.

We respectfully request that you perform an evaluation of the source individual who was involved. Given the circumstances surrounding this event, please determine whether our employee is at risk for infection and/or requires medical follow-up.

Attached is a "Documentation and Identification of Source Individual" form, which was initiated by the exposed worker. Please complete the source individual section and communicate the findings to the designated medical provider.

The evaluation form has been developed to provide confidentiality assurances for the patient and the exposed worker concerning the nature of the exposure. Any communication regarding the findings is to be handled at the medical-provider level.

We understand that information relative to human immunodeficiency virus (HIV) and AIDS has specific protections under the law and cannot be disclosed or released without the written consent of the patient. It is further understood that disclosure obligates persons who receive such information to hold it confidential.

Thank you for your assistance in this very important matter.

Sincerely,

SAMPLE FORM: DOCUMENTATION AND IDENTIFICATION OF SOURCE INDIVIDUAL

Name of exposed employee: _____
Name and phone number of medical provider who should be contacted:
_____

Incident Information
Date: _____
Name or medical record number of the individual who is the source of the exposure:
_____

Nature of the Incident
_____ Contaminated needle stick injury
_____ Blood or body fluid splash onto mucous membrane or nonintact skin
Other:
_____

Report of Source Individual Evaluation
Chart review by: _____ Date: _____/_____/_____
Source individual unknown _____ Researched by: _____
Date: _____/_____/_____
Testing of source individual's blood. Consent: obtained _____ refused _____
Check One:
_____ Identification of source individual infeasible or prohibited by state or local law.
State why if infeasible: _____
_____ Evaluation of the source individual reflected no known exposure to bloodborne pathogens.
_____ Evaluation of the source individual reflected possible exposure to bloodborne pathogen and medical follow-up is recommended.
Person completing report: _____ Date: _____/_____/_____
[*Note:* Report the results of the source individual's blood tests to the medical provider named above who will inform the exposed employee. Do not report blood test findings to the employer.]
HIV-related information cannot be released without the written consent of the source individual.

SAMPLE FORM: EMPLOYEE EXPOSURE FOLLOW-UP RECORD

Employee's name: _____ Job title: _____
Occurrence date: _____/_____/_____ Reported date: _____/_____/_____
Occurrence time: _____
Source Individual Follow-Up
Request made to: _____
Date: _____/_____/_____ Time: _____
Employee Follow-Up
Employee's health file reviewed by: _____
Date: _____/_____/_____
Information given on source individual's blood test results:
Yes _____ Not obtained _____
Referred to health care professional with required information:
Name of health care professional: _____
By whom: _____ Date: _____/_____/_____
Blood sampling/testing offered:
By Whom: _____ Date: _____/_____/_____

Vaccination offered/recommended:
By Whom:_____ Date: _____/_____/_____
Counseling provided:
By Whom:_____ Date: _____/_____/_____
Employee advised of need for further evaluation of medical condition:
By Whom:_____ Date: _____/_____/_____

[This questionnaire assists in evaluating and contracting for a transport, handling, and disposal company, should a facility not be equipped to handle its regulated waste. To avoid future liabilities from the inappropriate transport, handling, or disposal of regulated waste, it is imperative to identify legitimate contractors and document proof of the proper and legal transport, handling, and disposal of waste. Always keep the manifests that accompany shipments, and ensure that the waste has been transported and disposed of properly.]

SAMPLE CHECKLIST: REGULATED WASTE CONTRACTING

Request the company's identification number _____
Request to review the manner of recordkeeping _____
Documentation to include: _____
• List of items collected _____
• Method of destruction _____
• Site for destruction _____
• Proof of destruction _____
Requested information on insurance and bonding:
_____
_____
_____

SAMPLE CHECKLIST: POSTEXPOSURE FOLLOW-UP

Definitions:

1. **Exposure Incident:** A specific eye, mouth, other mucous membrane, nonintact skin, or parenteral contact with *blood,* or *other potentially infectious materials,* such as urine, feces, sweat and tears with visible blood, *that results* from the performance of an employee's duties.
2. **Universal/Standard Precautions:** All employees must utilize Universal/Standard Precautions, as an infection control method, which requires employees *to assume that all human blood and other potentially infectious materials,* such as urine, waste, sweat, or tears with visible blood, are infectious for HIV, HBV, HCV, and other bloodborne pathogens, and *must be treated accordingly.*

Introduction:

Following an exposure incident the employer must immediately make available to the affected employee a confidential medical evaluation and follow-up within the crucial 2-hour time frame. These procedures following an exposure incident may be audited by completing the following checklist.

1. Did you document the route(s) of exposure?   Y/N
2. Did an investigation occur that evaluated the circumstances under which the exposure occurred?   Y/N

Does this documentation include:
- Volume of blood or urine, feces, sweat, tears with visible blood injected into your body?   Y/N
- Length of time that any bodily substance was injected into the exposed employee's body?   Y/N
- Depth of the penetration from the contaminated sharp?   Y/N
3. Were investigative measures taken to determine the source of exposure?   Y/N
4. Was the source individual determined?   Y/N
   - If not, is there documentation stating why this was not feasible?   Y/N
5. Is it known whether the source individual is infected with HCV, HBV, and/or HIV?   Y/N
6. If it is not known whether or not the source individual is infected with HCV, HBV, and/or HIV, was permission obtained to examine his or her blood for baseline testing?   Y/N
7. Was permission given to test the source individual's blood further for possible infection to HCV, HBV, and/or HIV?   Y/N
   - If the source individual refused blood testing, was this documented?   Y/N
8. Was the affected employee informed of applicable disclosure laws and regulations concerning the source identity and infectious status?   Y/N
9. Was permission obtained to reveal the results of the blood test from the source individual to the affected employee?   Y/N
10. Was consent attained to collect the exposed employee's blood and test for HBV, HCV, and HIV serological status?   Y/N
11. If consent was attained, was the blood sent to an accredited laboratory?   Y/N
12. Was an identification number attained from the laboratory prior to sending the sample of blood, which complies with confidentiality requirements?   Y/N
13. If consent was not obtained to analyze the exposed employee's blood, has a sample of the blood been archived for the 90-day holding period?   Y/N
   - Was the exposed employee's refusal to have blood tested documented?   Y/N
   - If the exposed employee refused to have blood collected for archiving for the 90-day holding period, was this documented?   Y/N
14. Has the employee been referred to appropriate health care personnel to determine if postexposure prophylaxis (PEP) is recommended, as per Centers for Disease Control (CDC) guidelines?   Y/N
15. Did the PEP commence within 2 hours of exposure?   Y/N
16. Has the health care provider been supplied with:
   - A copy of the Bloodborne Pathogens Standard?   Y/N
   - A description of the exposed employee's duties, as they relate to the incident? Y/N
   - Documentation of the routes of exposure and circumstances under which the incident occurred?   Y/N
   - Results of the source individual's blood testing, if available?   Y/N
   - All medical records relevant to the appropriate treatment of the employee, including vaccination status?   Y/N
17. Has the health care provider submitted a written opinion to the employer within 15 days?   Y/N
18. Has the employer submitted the results to the exposed employee within 15 days?   Y/N
19. Has the employee received medical counseling on the medical evaluation results? Y/N
Were all of the following points discussed:
   - Exposed employee's risk of infection?   Y/N
   - Symptoms of HBV, HBC, and HIV disease?   Y/N
   - Availability of continued treatment?   Y/N
   - A description of the reevaluation phases to ensue, if exposed?   Y/N

20. Have factors related to the exposure incident been evaluated, including:
    - Evaluation of engineering controls and work practices surrounding the exposure incident?   Y/N
    - Evaluation of personal protective equipment used at the occurrence of the incident?   Y/N
    - Evaluation of policy and failures of associated controls at the time of the exposure incident?   Y/N
    - Evaluation of available medical technology, such as active safety features, to prevent future exposure incidents?   Y/N
21. Have all necessary forms been filled out completely, including:
    - An "Exposure Incident Report form"?   Y/N
    - A request for "Source Individual Evaluation"?   Y/N
    - An Employee Exposure Follow-Up Record"?   Y/N
22. Have all medical records documenting occupational exposure been retained for the course of employment plus for an additional 30 years?   Y/N

V. COMPRESSED GAS

A. NOTE

Compressed gas containers are to be stored in the following manner: Capped *and* secured to either a structural member (wall mounted, or support column), or on a cylinder stand, or cylinder truck. This is very simple. But its very simplicity makes for an easy citation, which would be difficult to argue, if *any* deviation from such procedures is found. Good protocol, which OSHA would like to see, is that the cylinders be tagged "Used" or "Empty," "In-Use," and "New."

B. BASIC RULES OF CYLINDER SAFETY

- Never use cylinders as table legs or to hold up other objects.
- Never hammer, pry, or wedge a stuck or frozen cylinder valve to loosen it, and never use a wrench. If a valve will not open by hand, call the gas distributor.
- Do not drop a cylinder.
- Do not allow grease, oil, or other combustible materials to touch any part of a cylinder. This rule is especially important when oxygen cylinders are involved. Grease or oil that oxidizes very slowly in air will burn in flames in pure oxygen.
- Never use a cylinder unless the gas it contains is clearly stenciled on it, or marked with a decal. Altering or defacing the name, numbers, or other markings is illegal.
- Do not rely on the color of a cylinder to identify the gas inside. Different suppliers use different color codes. Return an unidentifiable cylinder to the supplier.
- Keep cylinders away from electrical circuits and excessive heat (conducts electricity).
- Never strike an arc or tap a welding electrode on a cylinder.
- Keep cylinders away from sparks, etc. that result from welding, cutting, etc.
- Do not use or store cylinders where they may get hotter than 130°F.
- If a cylinder that has been stored outside is frozen to the ground, use only warm water to free it.

VI. CONFINED SPACE

A. INTRODUCTION

OSHA developed the Confined Space Standard to establish safety and health requirements for entry into *confined spaces*. The standard addresses hazards based on investigations into serious injuries and fatalities occurring in confined spaces. The types of hazards that are encountered in confined spaces include toxic and explosive atmospheres and mechanical hazards.

Long-term care facilities traditionally have not considered themselves to be in a high-hazard industry. In essence it really is not, compared with construction, trucking, mining, manufacturing, etc. The attention showered upon it now by OSHA is a product of its high incident rate of nonfatal injuries. This is driven mostly by the very real and poor industry history of back injuries in patient transfers. This industry is also the fastest-growing sector of employment. These reasons made the industry a priority of the OSHA agenda.

Most long-term care operators would look upon the Confined Space Standard and immediately dismiss it as irrelevant to their situation. Naturally, few laypeople would readily think that there are any confined spaces, or at least any of risk of explosion, suffocation, toxic exposure, or mechanical hazards, in a long-term care facility. However, there are certain areas/where the standard does apply to long-term care facilities, including elevator pits, sump pump rooms, closed rooms with pipe drainage, and septic or wastewater systems, to list the more common sites. Each employer is required to survey the workplace and *identify* each "confined space." Once the confined spaces have been identified, employers must *determine* if the space is a *Non-Permit-Required Confined Space* or a *Permit-Required Confined Space* and manage them accordingly.

The key is to understand that confined spaces by definition can include situations beyond ordinary preconceptions of what a confined space is. As you peruse this document, note the defining qualities of a confined space [*Note:* This guide has been developed specifically for the unique scenarios in long-term care facilities. It assists in identifying the confined spaces, and rendering them "Non-Permit-Required Confined Spaces" so that the facility will not be burdened by the requirements of a "Permit-Required Confined Space." If a facility finds that it indeed has a confined space that cannot be rendered a Non-Permit-Required Confined Space, seek professional safety assistance, as the materials necessary to cope with such a space are not specifically covered in this book.]

B.   Definitions

A confined space has the following attributes:

1.   An area that is large enough and so configured that an employee can bodily enter and perform assigned work.
2.   An area that has limited or restricted means for entry or exit (for example, water tanks, waste tanks, vessels, storage bins, hoppers, vaults, elevator pits, other pits, etc.).
3.   An area that is not designed for continuous human occupancy.

A non-permit-required-confined space is a confined space, as defined above, that *does not contain* any atmospheric or mechanical hazards that may result in serious injury or death. Thus, it does not require a permit for entry.

Additional terms to determine if a space is a permit- or non-permit-required space:

Lower explosive limit (LEL) is the minimum concentration of vapor in air below which propagation of a flame does not occur in the presence of an ignition source. The air-to-fuel ratio is *too lean for ignition* to occur below the LEL.

Upper explosive limit (UEL) is the maximum concentration of flammable vapor in air above which propagation of a flame does not occur on contact with an ignition source. The air-to-fuel ratio is *too rich for ignition* to occur above the UEL.

An oxygen-deficient atmosphere is an atmosphere having an oxygen concentration of less than 19.5% by volume. A person entering an oxygen-deficient atmosphere may die from lack of oxygen (asphyxia). (This may occur when oxygen is displaced by some other gas, usually a heavier gas that has settled into an enclosed space without adequate ventilation.)

An *oxygen-enriched atmosphere* is an atmosphere that contains more than 22% oxygen by volume. Open flames and ignition sources burn more intensely in oxygen-enriched atmospheres.

Permissible exposure limit (PEL) is the airborne chemical exposure limit that is published and enforced by OSHA for a number of chemicals.

Permit-required confined space is defined as a confined space that contains, *or has the potential to contain,* a hazardous atmosphere. A *hazardous atmosphere* may expose employees to the risk of death, incapacitation, impairment of the ability to escape unaided from a confined space, injury or acute illness from one or more of the following causes:

- Flammable gas, vapor, or mist in excess of 10% of its lower flammable limit (LEL, as defined above);
- Airborne combustible dust at a concentration that meets or exceeds its LEL;
- Atmospheric oxygen concentration below 19.5% or above 23.5%;
- Atmospheric concentration of any substance for which a dose or a PEL is listed by OSHA that could result in employee exposure in excess of its dose or PEL;
- Any other atmospheric condition that is immediately dangerous to life or health;
- A material that has the potential for *engulfing* an entrant; engulfment is the surrounding and effective capture of a person by a liquid or finely divided (flowable) solid substance that can be aspirated (inhaled) to cause death by filling or plugging the respiratory system, or that can exert enough force on the body to cause death by strangulation, constriction, or crushing;
- An internal configuration such that an entrant could be *trapped* or *asphyxiated* by inwardly converging walls or by a floor that slopes downward and tapers to a smaller cross section;
- Any other recognized safety hazard; the safety hazards can include an *electrical hazard, moving gears,* or a *rotating shaft,* to name a few.

A *confined space entry permit* is a document that is provided by the employer to allow and control entry into a permit space. It contains information specific to the confined space and hazards that are to be encountered upon entry. Permit-required confined spaces require such documentation. Permit-required confined spaces also require two-way instantaneous communications, a buddy system, and an on-site rescue emergency system.

C. FACILITY SURVEY AND HAZARD EVALUATION

1. Identify Confined Spaces

Survey the entire facility for confined spaces. Inventory all boilers, storage vessels, furnaces, tanks, manholes, sumps, open pits, diked areas, lagoons, sewers, tunnels, underground vaults, and elevator pits. Confined spaces come in many shapes and sizes. A permit-required confined space may not have an entry port. Some may be easy to recognize, like a storage tank, while others may not be so easy to identify, such as open pits, diked areas, or lagoons. Spaces such as an open-top water tank or sumps have walls that may restrict the movement or air, which may allow gases and vapors heavier than air to accumulate and displace oxygen. Be wary of any area that has characteristics that may restrict the movement of air; pockets of hazardous gases may accumulate. It is always safer to assume an unknown space is a permit-required confined space.

2. Hazard Evaluation

Determine the type of condition that may be associated with the confined space. The hazard may result in an atmospheric and/or mechanical hazard.

a. Atmospheric hazards

The types of gases that may create an atmospheric hazard include the following:

Hydrogen Sulfide

Hydrogen sulfide is a toxic, colorless, combustible gas that is heavier than air. It is formed by the decomposition of organic plant and animal life by bacteria. Hydrogen sulfide poisons a person by building up in the bloodstream. The gas paralyzes the nerve centers in the brain that control breathing. As a result, the lungs are unable to function and the person is asphyxiated.

Hydrogen sulfide is easily detected by a strong *rotten egg* odor in low concentrations. However, relying on this odor to warn of the presence of hydrogen sulfide can be dangerous. High concentrations can rapidly paralyze the sense of smell. Even low concentrations desensitize the olfactory nerves to the point that an individual may fail to smell the presence of the gas even if the concentration of the gas suddenly increases.

Oxygen Deficiency

Normal air contains 20.9% oxygen. When the oxygen concentration falls below 19.5% the air is considered oxygen deficient. Entering an oxygen-deficient atmosphere will lead to asphyxiation. Oxygen deficiency can be caused by the following:

Combustion—welding and cutting torches, boilers, burners
Decomposition of organic matter—rotting food, plant, and animal life
Oxidation of metals—rusting

Example: The sump in the basement of the building is connected to the sewer or cesspool system. Any gases that are generated by the decomposition process can seep into the sump via the piping. Inside the sump, the gases can accumulate and *displace the available oxygen* or *create an explosive atmosphere.*

Carbon Dioxide

Carbon dioxide is a colorless, odorless, and noncombustible gas. Elevated levels in the atmosphere will result in asphyxiation. Carbon dioxide is heavier than air and will settle in depressions such as pits and displace the available oxygen. This will result in an oxygen-deficient atmosphere. Carbon dioxide is generated by animal and plant respiration, organic decay, and fermentation.

Example: Potential areas of concern include but are not limited to unventilated fumes from boilers allowed to accumulate in a confined space.

Propane Gas

Propane is used as a fuel source for mechanical equipment such as forklifts. Propane is heavier than air and will settle in areas of lower elevation. The gas is both colorless and odorless. A foul-smelling odorant is often added when propane is used as fuel. Propane can displace oxygen, resulting in asphyxia. Propane is also a flammable gas that can ignite or explode in the presence of an ignition source.

Example: Building and storage configuration is such that propane is stored above an underground space with little to no ventilation. The fuel is heavier than air, and, should there be a propane leak, it could flow into that confined space and create a hazardous atmosphere. First, it can create an explosive atmosphere when the concentration of fuel to air is optimal for ignition. Second, after passing that concentration level, it can create a toxic atmosphere by displacing all the oxygen.

b. Mechanical hazards

Confined spaces, because of their often tight configuration, can place workers in dangerous proximity to physical and electrical hazards. Mechanical hazards that are associated with confined spaces include rotating shafts, agitators, raw material, live steam, movable platforms, rotating gears, and moving blades. A confined space entry permit is required to enter such confined space.

Example: The area at the base of the elevator shaft is a confined space. Anyone entering the base of the shaft is exposed to the hazard of being crushed by the descending elevator car. To render this space a non-permit-required confined space and merely a confined space, implement Lockout/Tagout to remove the mechanical hazard whenever work is to be performed.

3. Employee Exposure Assessment

Determine the type of work (scheduled maintenance, emergency repairs, opening or closing of valves, removing debris and cleaning) that will require the employee to enter the confined space. What is the frequency of entry into the confined space? Daily? Weekly? Monthly? Annually? In emergencies?

4. Classification of Confined Space

Once each confined space has been identified, determine if the confined space is a permit-required confined space or a non-permit-required confined space. If a confined space has, or has the potential to contain, a hazardous atmosphere, it *must* be classified a permit-required confined space. Entry can occur only *after* a confined space entry permit is completed. In most cases, only a confined space entry program is needed. In addition, OSHA requires a confined space entry permit prior to entry into permit-required confined spaces.

If a confined space has mechanical equipment that can create a mechanical hazard, the confined space is classified as a permit-required confined space. This usually applies to elevator rooms, pits, shafts, HVAC systems, etc. However, there are two options available to render these spaces non-permit-required confined spaces. First, if possible, guard the moving parts so that they do not pose a mechanical hazard. If, however, this is not practical, implement Lockout/Tagout whenever work is to be done in these areas. That will remove the mechanical hazard and the space will therefore not be defined as a permit required confined space, but a confined space only.

One can classify a confined space as a non-permit-required confined space *if the potential for an atmospheric hazard is found not to exist and/or the mechanical hazard is controlled.* Please refer to Section VI.C2a for managing the atmospheric hazard and Section VI.C2b for the mechanical hazard. A sample facility survey form follows.

SAMPLE SURVEY: CONFINED SPACE FACILITY

Date of survey: _____

Survey conducted by: _____

Location of Confined Space Identified	Type of Hazard	Permit Required?
_____	_____	_____
_____	_____	_____
_____	_____	_____
_____	_____	_____
_____	_____	_____
_____	_____	_____
_____	_____	_____
_____	_____	_____

D. Reclassification of a Potential Permit-Required Confined Space into a Non-Permit-Required Confined Space by Determining the Airborne Contaminant Concentration

This procedure is for confined spaces that are below grade level, such as pits, vaults, manholes. It is designed to assist in the classification of a confined space as a non-permit-required confined space, whenever possible.

Conduct a survey to identify all chemicals that are being stored, used, or generated in the area surrounding the confined space. Check for leaking containers and spills. Refer to the material safety data sheets (MSDS) to determine the physical properties of the chemicals. Chemicals that are heavier than air settle in areas of lower elevations (example: propane is a heavier-than-air chemical that may be used in the facility). One will need to conduct air monitoring to determine the level of these chemicals within the confined space.

If the confined space is connected to the sewer or cesspool systems, determine the sources of contaminants that may seep into the space. The toxic gases that are being generated within the sewer, cesspool, and associated piping may collect within the confined space.

Please refer to Section VI.C2a for a list of typical atmospheric hazards and examples. Once these chemicals are identified and their potential to seep into the confined space is determined, an air-sampling strategy can be devised. Refer to Section VI.D, air sampling worksheet, which will assist in devising a sampling strategy. When the atmospheric content within the confined space is determined and no atmospheric hazard is found to exist, then the space can be classified as a non-permit-required confined space.

Sampling Equipment

Equipment that can be used to determine the airborne concentration within the confined space can be ordered from any reputable safety supply company, i.e., Lab Safety Supply Company 800-356-0783. Order Diffusion Detector tubes manufactured by Drager or Dosimeter Tubes manufactured by Sensidyne. These tubes accurately measure airborne levels of specific gases. One will need to determine in advance the type of airborne contaminant likely to be encountered within the confined space.

The sampling tubes should broken at both ends upon immediate use, and be placed in the confined space for a sampling time of at least 4 hours. A color change will occur within the tube. The color change is the result of reactions caused by the gas in question with the chemicals in the tube. This is known as a colormetric reaction. The colormetric reaction will proceed along the length of the tube. Follow the manufacturer's instructions when using the colormetric tubes.

Refer to publications such as the American Conference of Governmental Industrial Hygienists (ACGIH) Threshold Limit Booklet, the NIOSH Pocket Guide to Hazardous Chemicals, and the OSHA Permissible Exposure Levels 1910.1000. These documents list the legal and recommended exposure limits for chemicals. It may be advisable to seek professional assistance in determining the exposure limits. This information should also be on the MSDS of the substances sampled, although MSDSs are notorious for inaccuracies.

The sampling should be conducted periodically to collect data from all four seasons and varying meteorological conditions. *It is recommended that the sampling be conducted over a 1-year period, during different meteorological conditions.*

When air sampling consistently determines that the levels of airborne contaminants are below the levels specified in the ACGIH Threshold Limit Booklet, the NIOSH Pocket Guide to Hazardous Chemicals, and the OSHA Permissible Exposure Levels 1910.1000, the confined space can be reclassified from a permit-required confined space to non-permit-required confined space. All records that support the reclassification must be retained.

SAMPLE WORKSHEET: AIR SAMPLING

Date: _____/_____/_____
Time: _____
Person Sampling: _____
Material or chemical sampled for: _____
Manufacturer or sampling tube: _____
Location of confined space: _____
Location of sample: top _____ middle _____ bottom _____ of confined space
Outside temperature: _____
Weather: _____
Snow cover: yes _____ no _____
Ground Frozen: yes _____ no _____
PEL for contaminant: _____
Sample end time: _____
Sample start time: _____
Sample total time: _____ (in minutes)
Concentration of contaminant: _____ (parts per million, ppm)
(Read the number from the scale on the dosimeter tube)
Time-weighted average exposure: _____ (ppm)
(Divide the concentration of contaminant by the total time in minutes of sample)

E. RECLASSIFICATION OF A PERMIT-REQUIRED CONFINED SPACE TO A NON-PERMIT-REQUIRED CONFINED SPACE BY REMOVAL OF SAFETY HAZARDS WITH THE IMPLEMENTATION OF LOCKOUT PROCEDURES

If the confined space is classified as a permit-required confined space solely because of the existence of a safety hazard created by mechanical equipment (pinch point, crushing), the application of a Lockout/Tagout procedure to prevent the unexpected energizing or start up of the machine or equipment reclassifies the permit-required confined space to a non-permit-required confined space. Only after the equipment is locked and tagged out, can the confined space be reclassified and employees allowed to enter safely and perform the assigned task.

Example: The pit of an elevator shaft is classified as a permit-required confined space. The mechanical hazard is the movable elevator car, which can easily crush an employee.

The application of a Lockout/Tagout procedure to reclassify the confined space must be certified. The person who installs the lockout and tagout equipment is the certifying individual. The certification must include the date, location of the confined space, and the signature of the certifying individual. A form that can be used to reclassify the permit-required confined space follows.

SAMPLE CERTIFICATION DOCUMENT—RECLASSIFICATION OF A PERMIT-REQUIRED CONFINED SPACE BY INSTITUTING LOCKOUT PROCEDURES

Date: _____/_____/_____
Location of confined space: _____
Method used to remove safety hazard: _____
Certifying individual: _____

F. EMPLOYEE TRAINING

The *affected employees* who may enter and work in a confined space must receive training that will impart the understanding, knowledge, and skills necessary for the safe performance of the duties that are required when entering a confined space. Training must include:

- Awareness of the location of each confined space.
- The type of hazard (if any) that is associated with the specific confined space.
- The procedures that are to be followed prior to entering the confined space.
- Training in hazard recognition, and the mode, signs, symptoms, and consequences of over-exposure to hazardous atmospheres.
- The purpose, proper use, and limitations of personal protective equipment, rescue equipment, and the tools used in confined space entry.

Records of employee training should be kept. A sample record-keeping document follows.

SAMPLE DOCUMENT: TRAINING RECORD-KEEPING

Date of Training: _____   Training Conducted by: _____

Name of Employee	Department	Signature
_____	_____	_____
_____	_____	_____
_____	_____	_____
_____	_____	_____
_____	_____	_____
_____	_____	_____
_____	_____	_____
_____	_____	_____
_____	_____	_____
_____	_____	_____

G. SAMPLE POLICY: CONFINED SPACE ENTRY

Definition of a Confined Space

It is an area that is large enough and so configured that an employee can bodily enter and perform assigned work. The work may include maintenance or repairs of equipment installed in the confined space, has limited or restricted means for entry or exit (for example, tanks, vessels, silos, storage bins, hoppers, vaults, and pits), and is not designed for continuous human occupancy.

Purpose

This document provides guidelines to define and identify confined spaces, and protocols to be followed when entering confined spaces.

Location of Confined Spaces

A survey of the entire facility is conducted. [See Section VI.C for guidelines and sample document.] Each location that meets the definition of a confined space is identified. The type of hazard, potential or actual, associated with the confined space should be determined. The hazard can include atmospheric, mechanical, or a combination of both. This information is recorded within the policy. [Sample document of findings:] As a result of a survey conducted in this facility the [example: pit, sump, vault] located [identify location] has been designated as a confined space. The type of hazard that may be encountered when entering this confined space is [atmospheric, mechanical, or a combination of both].

Hazards Associated with Confined Space

Atmospheric hazards

A confined space may have direct access to the sewer or cesspool system. This direct access may permit the accumulation, within the confined space, of toxic gases that are generated by the decomposition of organic material. The toxic gases may seep into the confined space and accumulate. The toxic gases may incapacitate or kill an employee(s) who enters the confined space.

Mechanical hazards

A confined space may house mechanical equipment. Examples of this type of confined spaces may be the base of an elevator shaft, or a machine pit for compacting equipment. Employees who must enter the confined space to perform repairs or maintenance may come into close proximity of rotating or moving equipment. The rotating or moving equipment may result in serious injury or death to the employee.

Procedure

[*Reminder:* The requirement of the OSHA Confined Space Standard is to identify all confined spaces and determine if the space is a permit-required confined space. If the space is a permit-required confined space, then specific protocols, equipment, and procedures must be in place before a confined space can be entered.]

Air monitoring must be conducted in confined space(s) with potential atmospheric hazards, to determine if an actual atmospheric hazard exists. The air monitoring is specific for the type of air contaminant that is suspected. The results of the air monitoring for *each* confined space is available for the employees to review prior to entering the confined space.

When air monitoring determines that an atmospheric hazard does not exist in the confined space, the confined space is classified as a non-permit-required confined space. Entry into this confined space without an entry permit is then allowed. A non-permit-required confined space does not fall under the scope of the OSHA confined space standard. [Please refer to Section VI.D: Reclassification.]

If the only hazard associated with a confined space is a mechanical hazard, the permit-required confined space can be reclassified a non-permit-required confined space by using a Lockout/Tagout procedure to isolate the mechanical equipment, and prevent unintentional activation of the equipment. If Lockout/Tagout procedures cannot be implemented, a confined space entry permit will be required prior to entering the confined space. [Please refer to Section VI.E: Removal of Safety Hazards.]

Training

The affected employees who may enter a confined space should receive training that will impart the understanding, knowledge, and skills necessary for the safe performance of the duties that are required when entering a confined space. The employee should be aware of the location of each confined space; the type of hazard (if any) that is associated with the specific confined space; and the procedures that are to be followed prior to entering the confined space. The designations of non-permit-confined space and permit-confined spaces are reviewed, whether entry into a confined space requires an entry permit, and the location and results of air testing that was conducted to determine the airborne hazard associated with a particular confined space. The training should be certified by recording the employee's name, a summary of the topics covered by the training, the identity and signature of the trainer(s), and the date training was conducted.

Contractors

All contractors are made aware of the potential hazards at the work site including the hazards involved in confined spaces. Contractors are to work equipped with the necessary elements to work in the confined space, or permit-required confined space, including but not limited to air monitoring, lockout/tagout, additional continuous ventilation, two-way instantaneous communication, and on-site emergency rescue when warranted.

VII. ELECTRICAL SAFETY

A. INTRODUCTION

There are a number of electric safety standards. Again, as in machine guarding violations (see Chapter 3, Section XIII), electrical safety violations may not be the most relevant or serious safety problem in a facility, or may not affect workers' compensation premiums. However, inspectors will seek them out.

B. DISCUSSION

Some typical examples of potential violations include:

- Any use of extension cords, except "temporary use," i.e., seasonal holiday lights and power tools. Hardwiring or permanent electrical wiring should be installed if necessary. This is also a DOH requirement/issue.
- Frayed wires anywhere and in all appliances.
- Missing ground pins in all appliances.
- Electrical boxes with exposed wires.
- Electrical panels with exposed wires.
- Missing outlet covers.
- Missing or improper blanks or knockouts in electrical panels.

Low-voltage alarm or phone wires do not fall into this category. Ground fault interrupters (GFIs) should also be installed in bathrooms and kitchen areas near sinks, tubs, or other water sources (within 6 ft).

Electrical standards, like machine guarding violations, are old faithful standards that have withstood the test of time. They are safe and familiar grounds for the inspector. Such citations are free from controversy and "exotic" issues, unfamiliar to the manufacturing plant–minded inspectors and their supervisors.

C. SUMMARY

It would behoove facility operators to conduct a comprehensive wall-to-wall electrical safety survey to identify and correct any of these violations. There should be an annual appliance electrical integrity check as part of the preventative maintenance program. Colored tags denoting year inspected could be used. Electrical safety check stickers are available from safety product catalogs. They tend to be outrageously expensive. They can be easily made in-house, customized no less, on colored address labels. The label should denote the name of the facility, month, year, and a place for the checker to initial.

VIII. EMERGENCY PLAN

A. DISCUSSION

Most facilities have elaborate emergency and disaster plans, replete with details addressing resident safety and the continued operations of the plant, but none of the few elements that OSHA requires. Ironically, the OSHA requirements could fit on a single page. An emergency plan under OSHA must include employee training on evacuation routes, an official meeting place, and a system of employee census after evacuation.

Sometimes, misunderstandings arise since most facilities do not ever really plan on evacuating the building, except in the very worst-case scenarios. Evacuations are conducted horizontally and vertically within the building. You can explain this matter of semantics to the inspector that evacuations are conducted in this manner first, and then full building evacuations afterward as necessary.

But also consider and write into the plan the following: What would happen after a full evacuation? How are employees to leave? Where are employees to meet? How will all the employees that were on site be accounted for? How will employees not on the site be accounted for? This last question mandates a good census system, not only to identify missing persons, but also to prevent emergency personnel from risking their lives by entering a dangerous building to save someone who was not there in the first place.

If there is annual fire extinguisher training, then it is theoretically not necessary to have an evacuation plan (see Section X). On the other hand, OSHA does not require that fire extinguisher training be conducted if there is an adequate emergency evacuation plan. It is one or the other. Train to evacuate properly or train to stay and fight the fire properly. Of course, it is best to have both.

B. SUMMARY

An emergency plan must include:

- Employee training on evacuation routes
- Instruction on where to meet immediately after evacuation
- A system of employee census

IX. OSHA EYEWASH AND EYE PROTECTION

A. GOAL

To prevent workplace injuries and exposures to the eye, employers are required to provide appropriate eye protection from both physical, chemical, and biological exposures. To reduce the severity of injuries and potential illnesses as a result of corrosive liquids splashed to the eye, objects lodged into the eye, or pathogens entering the eye membrane, appropriate eyewashes are to be provided within the work area for immediate emergency use.

B. SUMMARY

Appropriate eye protection shall be furnished, "where there is a reasonable probability of injury." This is vague and, thus, discretionary. Review MSDSs. *If a product calls for the use of an eyewash* (i.e., most oven cleaners) then it is most likely caustic, or poses some other serious potential injury, *and it is mandatory to install an eyewash in the work area where the product is used.* The reasonableness of the probability of an injury in a given situation may be disputable among managers, unions, employees, inspectors, judges, and attorneys. Use prudent judgment and clear documentation on how circumstances requiring eye protection and eyewashes were determined. Note, if eye goggles are provided, there should be a policy and enforcement to support their *consistent* use.

Eyewashes are provided for *immediate emergency use,* as a form of first aid, within the work area in question. OSHA refers to ANSI (American National Standards Institute) to quantify the meaning of this. Highlights of ANSI standards Z3568.1-1998 are

- Eyewashes should be within 10 seconds travel distance unless the hazard involves strong acids or corrosives in which case it should be at the area of exposure.
- The path from the work area with potential exposure to the eyewash shall not be obstructed by a locked door.
- Eyewashes should be located in a well-lit area and identified with a sign.
- The water must be able to flow for a minimum for 15 minutes continuously.
- The water from an eyewash fountain should flow at a rate of 0.4 gallons per minute.
- Valves must activate within 1 second or less.
- Valves must stay open (leaving hands free).
- Eyewashes shall be installed within 33 to 45 in. from the floor.
- Eyewashes should be positioned 6 in from a wall or the nearest obstruction.
- Eyewash units shall be activated weekly to verify proper operations.
- Staff must be trained in the use of the equipment.

Thus, to ensure compliance, purchase and install only ANSI-approved eyewashes. (Also, do not confuse eyewashes with showers. There are combination eyewash/showers and showers, some of which contain only an ANSI-approved shower.) Handheld drench hoses support shower and eyewash units but shall not replace them.

C. UPDATE

OSHA updated its Personal Protective Equipment Standard (which includes eyewashes) in the summer of 1994. It enhances the existing standard by requiring employers to conduct a hazard assessment to determine the appropriate personal protective equipment needed, and to ensure that employees are provided with, and are using the appropriate equipment. This Standard revision also requires appropriate training on the proper use of the equipment (see Section XVI).

Nonplumbed single-wall units containing 20 oz saline-buffered antibacterial eyewash bottles are an inexpensive way to show "good faith" to OSHA, as well as concern for employees. These may be strategically placed in housekeeping carts or in certain areas to assist a potentially injured employee to move from point A to point B. *Eyewash bottles do not serve as eyewashes unless they are able to deliver 15 minutes of continuous flow at 0.4 gallons per minute.* They supplement plumbed installations. This refers to the section of MSDSs on eye exposure, which instruct the employee to flush with copious amounts of water for at least 15 minutes.

If an eyewash fixture is used on existing plumbing, it would be prudent to develop and enforce a policy that requires the nozzle to be switched onto the "eyewash" mode at all times, if possible, except for when someone is using the tap for regular faucet use. This may not be possible, as many eyewash manufacturers have designed the unit to automatically default back to regular tap use when the eyewash is not engaged.

If at all possible, shut off the hot water supply to this faucet, unless hot water is used at this faucet. If hot water is not disconnected, employees shall be instructed and expected to flush with cold water for a few seconds, after the use of hot water. People should get into the habit of these procedures. The more strictly these rules are followed, the better the defense is for the installation as adequate and posing no additional risk. Otherwise, inspectors may argue for the need for a dedicated eyewash.

X. FIRE EXTINGUISHER USE

A. DISCUSSION

It may not be optimal to train employees in long-term care facilities in the use of fire extinguishers, or encourage their use. Many fatalities, serious injuries, and worsening fires have resulted from employees attempting to put out fires with fire extinguishers when they would have been better off evacuating and confining the fire. Still, some facilities deem it prudent to provide the training with stern instruction on its conservative use and always to err on the side of safety first.

OSHA does not require that fire extinguisher training to be conducted per se *if a facility has an adequate emergency evacuation plan* (see Section VIII).

B. SUMMARY

If a facility plans on conducting fire extinguisher training, then it should ensure that employees are trained annually and that there is documentation to that effect. Training does not have to be hands-on, although that would be preferable for practical purposes. If a facility does not plan to conduct fire extinguisher training, then two items must be present:

- An emergency plan that includes employee training on evacuation routes, a formal meeting place, and a system of employee census.
- A formal policy of "no fire extinguisher use," and documented training that employees not use the fire extinguisher.

A problem arises when an inspector asks an employee if he or she would use a fire extinguisher if there was a little fire: the employee answers "Well, yeah, I guess" and then answers, "Well, no, not really" to the question, "Have you received fire extinguisher training?" Then there is a potential for a citation, with good reason, actually, since this shows that employees are inclined to use fire extinguishers, but have not been trained on their proper use.

XI. HAZARD COMMUNICATION

A. GOAL

Employers shall have a comprehensive chemical management program to minimize the risk of employee exposure to chemicals in an effort to prevent injuries and illnesses resulting from both chronic (long-term) and acute (short-term) exposures.

B. SUMMARY

Employers are required to have a hazard communication program in the following form:

- A *written hazard communication policy* delineating how the program is developed, implemented, and maintained.
- Uniform *labeling system* of all containers throughout the facility. (The only exception is if the container will be used by only one employee for one shift, and is disposed of, and not stored, by the end of that shift.)
- A *master chemical inventory* list of current products used in the facility by brand name in alphabetical order.
- *MSDSs* of all nonsolid substances at the facility (including pharmaceuticals), in alphabetical order by product name available for any shift, accessible to all employees. The

MSDS books should include only relevant and current MSDSs for that department, with a master MSDS book in the main office containing all MSDSs throughout the facility, and a dead MSDS file with disuse dates marked for future reference.

* *Training* of employees on the facility's hazard communication program, job-specific chemical hazards, first aid, spill cleanups, etc. as required by the training requirements; yearly training suggested.

C. UPDATE

This standard was originally conceived to protect chemical workers and employees in heavy manufacturing environments. This is an OSHA favorite citation. Five of the top 25 most frequently cited OSHA citations have come under the Hazard Communication Standard in the past 10 years. Every individual deficiency can count as a separate violation, carrying separate penalties, even though they stem from the same standard. All labels and training materials must be relevant, correct, visible, and in English, even if your workforce does not speak or read English. (It is not a "right to understand" rule.)

Several years ago, the American Hospital Association wrote a formal letter to OSHA requesting official clarification on MSDSs for pharmaceuticals. The authoritative response was that MSDSs are required for all medications that are not in a solid form or may not be administered in solid form. (For example, all liquids, propellants, and solid drugs that may be ground into powder before administering.) The standard exempts employees' personal use medications. However, the cosmetics, personal care products, and drugs used in a long-term care facility within its operations are not for employees' personal use and, thus, are covered. This is in keeping with basic industrial hygiene protocol and the intent of the standard. No facility has been cited for lacking MSDSs for pharmaceuticals and medications. Last, there has been some excitement over the new revision to the standard. Nothing has changed except that OSHA is officially now accepting electronic MSDSs as long as they are made available to employees during their workshift.

[*Note:* Disused MSDSs should be filed away indefinitely (over 30 years) because they are considered to be medical and exposure records under the Access to Medical and Exposure Records Standard. Employees and former employees have the right to copies of MSDSs of the chemicals that they worked with. Such records are to be kept for 30 years after the termination of the employment relationship. Also, some state and local department of health laws (i.e., New York) require that if an employer ceases to exist within that period of time, it is supposed to forward these records to the state DOH.]

D. CONTENTS

An outline of the Hazard Communication Program Outline is presented in Section XI.E.

A sample Hazard Communication Policy is provided in Section XI.F. When the policy is completed, be sure that it is actually implemented.

Also included are a sample letter requesting MSDSs from suppliers who insist MSDSs are not required of them (provides an interim document trail for missing MSDSs), a sample chemical safety management review, and a sample checklist for hazard evaluating. At the end of Section 3.XI is an MSDS terms glossary and chemical safety primer for the lay user.

E. HAZARD COMMUNICATION PROGRAM OUTLINE

  I. Objective
    A. Inform and Train Employees on Existing Chemical Hazard
    B. Prevent Chronic and Accidental Chemical Exposures Arising out of Misinformation or the Lack of Information/Education
  II. Master Chemical List
    A. Alphabetical and Complete
    B. Updated

      C. Name Can Be Cross Referenced to MSDS
      D. Use Product Brand Name
  III. Labeling Systems and Other Warnings
      A. Shipping and Receiving Only to Accept Properly Labeled Shipments
      B. Labeling on All Containers
         1. Uniform Chemical Identity
         2. Appropriate Hazard Warning
         3. Name/Address/Phone of Responsible Party or Manufacturer
         4. Legible and in English
      C. Labels Not to Be Removed or Defaced
      D. Prominently Displayed on Containers and Work Areas
      E. Exempt Portable Containers for Single Employee One-Shift Transfer Use
  IV. Material Safety Data Sheets (MSDSs)
      A. English, Chemical, and Common Name
      B. Physical and Chemical Characteristics
      C. Physical and Health Hazards (signs and symptoms of exposure)
      D. Primary Route of Entry
      E. TLV/PEL/NTP/IARC/RTECS, etc., Indications If Available
      F. General Control, Emergency, and First Aid
      G. Dated
      H. Updated
   V. Employee Training
      A. Given at Initial Assignment or when New Hazard Introduced, New Employee, and/or Job Transfer
      B. Methods and Observations to Detect Presence or Release
      C. Routes of Exposure and Types of Exposure
      D. Possible Effects of Exposure Both Chronic and Acute
      E. Emergency Procedures
      F. The Hazard Communication Program
         1. MSDS
         2. Chemical Inventory
         3. Labeling System
         4. Written Policy

F. SAMPLE POLICY: HAZARD COMMUNICATION

The written program must include the specific methods that are used to achieve compliance with the requirements of the Hazard Communication Standard, 29 CFR 1910.1200. The specific methods described in this sample written program are for illustrative purposes, and other effective methods may be substituted to satisfy specific needs or practices, although elements not specifically labeled optional are required.

1. General

The purpose of this policy is to ensure that [facility name] is in compliance with the OSHA Hazard Communication Standard. The [title of person in charge of program] is the overall coordinator of the facility program acting as the representative of [facility name], who has overall responsibility for chemical safety management.

    In general, each employee in the facility will be informed of the content of the Hazard Communication Standard, the hazardous properties of chemicals they work with, and measures to protect themselves from these chemicals. This Hazard Communication Program sets forth the systems to ensure the safe handling, use, and disposal of all chemicals on site.

2. Chemical Inventory [Required]

The [title of person in charge of program] will maintain a list of all hazardous chemicals used in the facility, and update the list as necessary. The hazardous chemical inventory will be updated upon receipt of new chemicals at the facility, and the disuse of any existing chemical on the list. This list is kept in alphabetical order by product name. This chemical inventory in the facility is maintained at [place].

3. Chemical Inventory and Purchase Review [optional]

[Optional, not required but a good chemical safety management activity.]

A review of all chemicals as presently used at the facility is conducted initially, and periodically. A determination will be made to ensure that these products are indeed effective to achieve desired results to do the job, but also, to the extent practicable:

- Are compatible as much as possible with other existing products to reduce the potential for chemical reactions (i.e., bleach-only facility, or ammonia-only facility)
- Are low in toxicity and other harmful effects
- Allow for small and easy to use storage containers (size and design that prevent larger leaks, spills, and ergonomic strain)
- Require minimal storage amounts (reduces overall potential for leaks, spills, and reactions, and storage space)
- Provide users with good labeling and information system

SAMPLE TOOL: CHEMICAL SAFETY MANAGEMENT REVIEW

[Optional tool to conduct comprehensive chemical safety review to identify hazards and prioritize issues.]

1. List three or more of the most often used products or products that are used or stored in the *largest volumes:*

   _____   _____
   _____   _____
   _____   _____
   _____   _____

2. List the more *toxic* products that are used or stored in your department, regardless of the amounts actually used or stored (toxic, acutely toxic, carcinogen, etc.):

   _____   _____
   _____   _____
   _____   _____
   _____   _____

3. List the more *hazardous* products that are used or stored in your department, regardless of the amounts (flammable, corrosive, operational hazard, etc.):

   _____   _____
   _____   _____
   _____   _____
   _____   _____
   _____   _____

4. Describe the emergency procedures (first aid, spill control, etc.) for the products that you feel should be prioritized based upon the above three lists:

Product: _____
Procedures: _____
_____
_____
_____

Product: _____
Procedures: _____
_____
_____
_____

Product: _____
Procedures: _____
_____
_____
_____

Product: _____
Procedures: _____
_____
_____
_____

Product: _____
Procedures: _____
_____
_____
_____

Product: _____
Procedures: _____
_____
_____
_____

SAMPLE CHECKLIST: AREA/OPERATION HAZARD EVALUATION

Evaluate each **Work Area, Storage Area, Job Classification,** and **Operation** in a department for *potential chemical safety hazards.* Consider routine *and* nonroutine situations. Some methods of evaluation include:

- Reviewing job descriptions in detail
- Site inspection of areas
- Careful observations of procedures
- Consideration of foreseeable incidences
- Review of past accidents and "near-misses"
- Cause-tree analysis
- Safe and unsafe behavior analysis

- "What If" analysis—Consider unplanned events: understaffing, climatic, energy, infrastructure disasters; equipment failure; improper procedures; etc. and the means to prevent the incident from occurring, and to minimize its effects or severity in case it should occur.

List the associated protective equipment, work procedures, and measures necessary to prevent and control these potential or even common occurrences.

Identify Job/Area/Operation: _____

Potential Hazard: _____

Potential Causes: _____

Preventive Measures: _____

Control Measures: _____

Identify Job/Area/Operation: _____

Potential Hazard: _____

Potential Causes: _____

Preventive Measures: _____

Control Measures: _____

4. Material Safety Data Sheets (MSDSs) [Required]

The [title] will maintain an MSDS library on every substance on Chemical Inventory in the [place]. The MSDS will consist of a fully completed OSHA Form 174 or equivalent. The [title] will ensure that each work area maintains an MSDS for hazardous materials used in that area. MSDSs will be readily accessible to *all* employees during *all* shifts.

The [title] is responsible for acquiring and updating MSDSs. The [title] will review each MSDS for accuracy and completeness. All new procurements for the facility must be cleared by the [title]. Whenever possible, the least hazardous substance will be used. MSDSs that meet the requirements of the Hazard Communication Standard must be fully completed and received at the facility either prior to or at the time of receipt of the first shipment of any potentially hazardous chemical purchased from a vendor. It may be necessary to discontinue procurement from vendors failing to provide approved MSDSs in a timely manner.

[Ensure that MSDSs are available for *all* departments, including the beauty salon, physical and occupational therapy, and recreation. Even if the person working with this material is not an employee, staff and residents can be exposed to them and MSDSs should be available to them for review, and in case of emergency. MSDSs are not necessary for any items that are labeled "nontoxic."]

5. Labels and Other Forms of Warnings [Required]

[Title or group title, i.e., supervisors] is designated to ensure that all hazardous chemicals in the [facility/their unit or area] are properly labeled. Labels should list at least the chemical identity, appropriate hazard warnings, and the name and address of the manufacturer, importer, or other responsible party. [Title] will refer to the corresponding MSDS to verify label information. Immediate-use containers, small containers into which materials are poured for use on one shift by an employee drawing the material and finished with by that employee within that shift, do not require labeling (i.e., a bucket). To meet the labeling requirement of the standard for other in-house containers, refer to the label supplied by the manufacturer. All labels for in-house containers will be approved by [title] prior to their use.

[Title] will check monthly to ensure that all containers in the facility are labeled and that the labels are up to date.

6. Training [Required]

Each employee who works with or who is potentially exposed to hazardous chemicals will receive initial training on the standard and the safe use of those hazardous chemicals. Additional training will be provided for employees whenever a new hazard is introduced into their work areas, or when they are transferred to a new area, and periodically (yearly). Hazardous chemical training is conducted by [trainer]. The training will emphasize the following elements:

- A summary of the standard and this written program
- Hazardous chemical properties including visual appearance and odor and methods that can be used to detect the presence or release of the chemicals
- Physical and health hazards associated with potential exposure to workplace chemicals
- Procedures to protect against hazards, i.e., personal protective equipment, work practices, and emergency procedures
- Hazardous chemical spill and leak procedures
- Where MSDSs are located, how to understand their content, and how employees may obtain and use appropriate hazard information

The [title] will monitor and maintain records on employee training and advise the facility manager on training needs.

Written policies and procedure [optional]

Before planning actual in-service, make sure that all the protective equipment is available; all the control equipment is available or installed (eyewashes, proper fire extinguishers, wet-vacs, etc.); all the necessary procedures developed from Phase 2 have been added or amended into existing procedures and policies. Make sure that all of these are congruent with existing policies, procedures, and practical operations.

Scope of training [optional]

Use the Hazard Communication Training Outline as a guide for the material to cover in staff training. Use it in conjunction with the information prioritized based on volume, toxicity, use, hazard, probability of occurrence, severity of result of occurrence despite low probability [see above for a sample chemical safety management review]. Hazard communication training would probably yield more meaningful results if staff is trained on proper work procedures, rather than strictly reviewing the MSDS with them, as much of the information on the MSDSs is rather technical and esoteric. A glossary of MSDS terms follows this sample policy. It would be best to provide these to supervisors and management personnel. It may help them develop a more-sophisticated comprehension of chemical safety. That would enable them to serve better in their capacity as supervisor to assess, plan, and organize for chemical hazards and risks, while relaying the information in lay terms for regular staff.

Delivery and presentation [optional]

Remember, reviewing the MSDSs will probably be too technical for staff, while a focus should be placed on practical applications (especially emergency procedures and first aid) and actual operational hazards.

Hazard communication training outline [required elements]
   I. Methods and Observations Used to Detect Presence/Release
      A. Chemical Monitoring
      B. Physical Evidence—Odors, Visual, Tactile
   II. Physical Hazards
      A. Explosivity
      B. Flammability
      C. Combustibility
      D. Reactivity

    E. Corrosivity
    F. Gases/Vapors/Particulates
III. Health Hazards
    A. Toxicity
    B. Acute Effects
        1. Irritant
        2. Acute Toxicity
    C. Chronic Effects
        1. Sensitizer
        2. Illnesses
    D. Local Effects
    E. Systemic Effects
    F. Routes of Exposure
        1. Inhalation
        2. Ingestion
        3. Absorption
        4. Injection
    G. Individual Susceptibility
IV. Protective Measures
    A. Proper Work Procedures
    B. Emergency Procedures
        1. Communication
        2. Cleanup
        3. First Aid
    C. Personal Protective Clothing and Equipment
V. Hazard Communication Program
    A. Written Policy
    B. Master Chemical Inventory
    C. Labeling System
    D. MSDSs

Scope of employees to be trained [required]

All staff should be trained. New staff must be trained during orientation before they work in a capacity that would expose them to the potential risks. All staff should be refreshed yearly. [OSHA
does not mandate that Hazard communication training be conducted on a yearly basis. However, due
to the turnover rate in this industry, and new DOH training requirements, hazard communication
training must be conducted for all staff on an annual basis.]

Training format

Facilities can train in seminar style or on a one-to-one basis. Be sure to document content, extent,
and dates of training, and obtain signatures of employees who have completed the training. It may
be beneficial to make up a quiz that can be administered before and after training to access incre-
mental advancement of employee knowledge.

7. Contractor Employers [Required]

The [title], upon notification from the [title of person in charge of contractor work], will advise outside contractors of any chemical hazards that may be encountered in the normal course of their work
on the premises. MSDSs from contractors will also be sought if employees may be exposed to anything they use on site.

8. Nonroutine Tasks [Required]

[*Titles of those possibly involved in nonroutine tasks*] contemplating a nonroutine task, for example boiler repair, will consult with the [title] and will ensure that employees are informed of chemical hazards associated with the performance of these tasks and appropriate protective measures. This will be accomplished by a meeting of supervisors and the safety manager with affected employees before such work is begun.

SAMPLE LETTER REQUESTING MSDS FROM SUPPLIERS WHO INSIST MSDSs ARE NOT REQUIRED OF THEM

Retain a copy of this letter in your files along with all correspondences received from the supplier to document your reasonable attempts to obtain the MSDS.

(Name)   (Date)

_____

(Title)

_____

(Name of Company)

_____

(Address)

_____

Re: Final Request for MSDS

Dear Mr/Mrs/Ms.[Name]:

We are in receipt of your correspondence dated [date of their letter] suggesting that your counsel [or an equally appropriate party] has advised that [Name of Company] is not required to provide MSDS on [Name of product or products in question].

It is our interpretation that the products which we are requesting MSDSs for are not exempt under Title 29 Code of Federal Register Part 1910.1200, the OSHA Hazard Communication Standard, in either section (b) Scope and Application, paragraphs (5) and (6), or section (f) Labels and Other Forms of Warning, paragraph (7), which describes exemptions.

A Products Development or similar department at your company may determine that the product in question is *nontoxic*. However, toxicity is only half the hazard equation. The degree of hazard a substance poses is dependent not only on inherent toxicity, but also on the manner in which it is used. This factor is more site specific. Different facilities use products in different environments and various ways that may pose more or less of a hazard to the degree that it has more or less of a potential to be misused, mishandled, mixed with other substances, etc.

Also, MSDSs serve as medical and health records of potential exposures. These are documents that employers are obligated to provide employees upon their request. This may be a defensive strategy for facilities to prove that the likelihood of a chronic or acute exposure is minimal, if not impossible, at their facility during the employee's tenure.

Thus, most prudent suppliers have provided MSDSs on the more seemingly innocuous products in order that facilities may complete their MSDS files, and continue to work with better knowledge of and more confidence in the products.

This will serve as our final request for the above. We thank you in advance for your anticipated cooperation.

Sincerely,

MSDS TERMS GLOSSARY AND CHEMICAL SAFETY PRIMER FOR THE LAY USER

CAS#: Chemical Abstract Number—universal reference for proper chemical identity.

Gases: Generally supplied in compressed gas cylinders, readily diffuse or spread out at room temperature.

Vapors: Formed by the evaporation of liquids or *solids*.

Particulate: Solid or liquid particles suspended or dispersed in air, examples: dust, mists, fumes, smoke, tubercle bacteria, etc.

Relative degree of severity of hazard in diminishing order: DANGER → WARNING → CAUTION

Flash point: Minimum temperature at which a liquid gives off a vapor in sufficient concentration to ignite (when tested following established flash point testing protocols).

Flammable: Any liquid having a flash point below 100°F.

Combustible: Any liquid having a flash point at or above 100°F *but* below 200°F.

Reactive: Substance that has the tendency to explode under normal conditions, to react violently when mixed with water, or generate toxic gases (unstable).

Water reactive: Reacts with water to release a gas that is flammable or presents a health hazard.

Flammable/explosive limits: Taken as a percent by volume of air, the range of concentration of the substance in the air when there is a hazard. Below the range, the concentration is too dilute; above the range, it is too thick (not enough oxygen mix).

Corrosive: Chemical that causes visible destruction of, or *irreversible alterations* in, living tissue by chemical action at the site of contact. This does *not* refer to inanimate surfaces.

Irritant: A chemical that is not corrosive, but which causes a *reversible inflammatory effect* on living tissue by chemical action at the site of contact (i.e., latex powder allergies, dermatitis from detergent use).

Explosive: Chemical causing sudden, almost instantaneous release of pressure, gas, and heat when subjected to sudden shock, pressure, or high temperatures.

Oxidizer: A chemical that initiates or promotes combustion in other materials, thereby causing fire either of itself or through the release of oxygen or other gases.

Organic peroxide: Organic compound that contains the bivalent O–O structure, which may be considered a derivative of hydrogen peroxide where one or more of the hydrogen atoms have been replaced by an organic radical (worse than oxidizer).

Dose: Combination of concentration of the substance and the length of time of exposure.

Toxicity: The capacity of a substance to produce an unwanted effect when that substance has reached sufficient concentration in the body. Usually described by the LD_{50}, or Lethal Dose 50%. This dose is the amount that when administered to laboratory animals kills half the population within 14 days. Thus, the lower the LD_{50} the more potent or toxic.

Toxic: 1. Oral LD_{50} of more than 50 mg/kg, but not more than 500 mg/kg. 2. An inhalation of LC_{50} (lethal concentration) or more than 200 ppm but not more than 2000 ppm of a gas or vapor, or more than 2 mg/l but not more than 20 mg/l of a dust, mist, or fume. 3. A dermal LD_{50} of more than 200 mg/kg but not more than 1000 mg/kg.

Acute toxicity: 1. Oral LD_{50} 50 mg/kg of body weight or less. 2. Inhalation LC_{50} of 200 mg/kg or less. 3. Dermal LD_{50} of 200 mg/kg or less.

Chronic effects: Effects suffered from chronic exposure—regular repeated exposure to a substance over a period of months, years, lifetime.

Acute effects: Effects suffered from instantaneous exposure, a short exposure usually minutes, hours, or several days.

Systemic effect: Damage to the body that is caused when a chemical is absorbed by the body and then acts on a specific organ. An example is liver damage caused by the inhalation of benzene (gas) or the ingestion of alcohol.

Local effect: Damage to those parts of the body that actually come in contact with the harmful substance. For example, the irritation of the throat and lungs produced by inhalation of hydrochloric acid or lung cancer from smoking.

Sensitizer: A chemical that causes a substantial proportion of exposed people/animals to develop an allergic reaction in normal tissue after repeated exposure.

Individual susceptibility: The sensitivity of an individual to a substance. The effects of exposure to a particular chemical are different for different people.

Carcinogen: A substance causing development of cancerous growths in living tissue.

Mutagen: A substance able to induce mutations in genes transmissible to offspring.

Teratogen: A substance that acts during pregnancy to produce a physical/functional defect in the fetus or offspring.

TLV: Threshold limit value set by ACGIH (American Conference of Governmental Industrial Hygienist). Not required by law, these are voluntary standards updated yearly.

PEL: Permissible exposure limit—mandatory standards set by OSHA for specific substances. Established in 1967, not updated since. Usually in terms of concentrations.

TWA: 8-hour time-weighted average based on a 40-hour workweek (average daily exposure taking into account exposure times of high and low exposure concentration).

C: Ceiling, not to be exceeded at *any* time, regardless of what the TWA might be.

STEL: Short-term exposure limit, the limit for a 15-minute exposure allowed only twice a day, only if no other exposure to that substance occurs that day.

SKIN: Indication that the substance is a skin irritant as determined by animal testing protocols.

Ca: Carcinogenic indication:

- If the substance is regulated by OSHA as a carcinogen
- If the substance is determined to be a confirmed or potential carcinogen by NTP (National Toxicity Program) or IARC (International Agency Registry of Carcinogens)

IARC: International Agency for Cancer—tests chemicals and rates them for toxicity, carcinogenicity, etc.

NTP: National Toxicology Program, domestic IARC.

RTECS: Registry of Toxic Effects of Chemicals—registry of all available health/toxicity information on a substance.

HMIS: Hazardous Materials Identification System for health, flammability, and reactivity rates from 0 to 4 (4 most hazardous).

NFPA: National Fire Protection Association, same as HMIS.

Examples of "Toxins"and Their Characteristics

Toxin type: Health effect (i.e., example of chemicals) the symptoms of exposure.

Cutaneous hazards: affecting the dermal layer of the body (i.e., ketones, chlorinated compounds), skin defatting, rashes, irritation, dermatitis.

Lung damaging agents: Irritating or damaging pulmonary tissues (i.e., silica, asbestos, ammonium chloride gas, mix of bleach and ammonia), cough, tightness in chest, shortness of breath.

Eye hazards: Affecting the eye or visual capacity (i.e., organic solvents, acids), conjunctivitis, corneal damage.

Hepatotoxins: Chemicals producing liver damage (i.e., carbon tetrachloride), jaundice, liver enlargement.

Nephrotoxins: Chemicals producing kidney damage (i.e., halogenated hydrocarbons, uranium), edema, proteinuria.

Neurotoxins: Substances producing primary effects on nervous system (i.e., mercury, carbon disulfide), narcosis, behavioral changes, decrease in motor function.

Hematotoxins: Chemicals acting on blood or hematopoietic system (i.e., carbon monoxide, cyanides), loss of consciousness, decreased hemoglobin function, deprives the body tissues of oxygen.

Reproductive toxins: Chemical affecting reproductive capabilities including chromosomal damage, mutations, and effects on fetuses (i.e., lead, DBCP), birth defects, sterility. These affect *both sexes.*

How safely the body handles a substance depends on:

- The *type* of substance.
- The *amount* absorbed.
- The *period of time* over which it was absorbed.
- The *susceptibility* of the person who is exposed.

Chemicals have a wide range of possible health effects. Many of these are good, but it is important to remember that too much of anything can be harmful. For example, aspirin in the correct amounts (dosage) can relieve minor pain and muscular aches, decrease the chances of a heart attack. But if too much aspirin is consumed, death can result. In addition, some people cannot take aspirin. *All substances are poisons—only the dose separates a poison from a remedy.*

XII. LOCKOUT/TAGOUT

A. GOAL

In an effort to prevent death and injuries due to inadvertent energizing of machines, steam or other systems, or circuits with potential energy, employers are required to implement a program to control hazardous energies.

B. SUMMARY

Employers must have a Lockout/Tagout policy, which governs the proper procedures for any electrical or mechanical work or repairs where the possibility of inadvertent start-ups or release of energy may occur. Employees are to be trained on the proper procedures to lock and tag switches, circuits, boxes, valves, etc. Employer shall conduct an annual inspection, and sign to certify that the annual inspection had been conducted, to ensure that the procedures are followed and the program is, in fact, implemented. Remedial training must be given and documented if deviations are found during the self-inspection by the employer.

C. UPDATE

This is a simple, sensible standard. Many employers have not complied because of a lack of awareness or a lack of desire to comply. It does not require much to comply. Simply make the locks and tags available, the policy and procedures clear, and inspect the health of the program yearly, and ensure employee compliance with the procedures.

Note: Make sure that your elevator extrication plan is adequate and followed. Lockout/Tagout applies to the need to deenergize the elevator, before any repair or extrication is to begin. The long-term care industry, and other nonmanufacturing service-oriented groups rarely experience any fatal or severe physical workers' compensation claims. Those few that do (regularly) occur have always involved improper elevators repairs or elevator extrication, and/or vehicular accidents.

D. CONTENTS

A sample written lockout/tagout program with a sample equipment evaluation procedures form, a sample training rosters form for both *authorized* (to do the work) and *affected* (all other) employees, and a sample annual inspection certification form.

E. SAMPLE PROGRAM: LOCKOUT-TAGOUT (HAZARDOUS ENERGY CONTROL PROGRAM)

1. Purpose

This procedure establishes the minimum requirements for the lockout and tagout of energy isolation devices. It shall be used to ensure that the machines or equipment are isolated from all potentially hazardous energy, and locked-out or tagged-out before employees perform any servicing or mainte-nance activities where the unexpected energization, start-up, or release of stored energy could cause injury.

Procedures have been established for isolating machines or equipment from the input of energy. Locks or tags are to be placed onto circuit breakers, disconnect switches, line valves, etc., which are the energy-isolating devices for the equipment being maintained. All equipment that can be locked out, must be locked out. The equipment that cannot be locked out must be tagged. The appropriate locks or tags are available to authorized staff members.

2. Preparation for Lockout/Tagout

Conduct a survey to locate and identify all isolating devices to be certain which switch, valve, or other energy-isolating device(s) apply to the equipment to be locked and tagged out. More than one energy source (electrical, mechanical, or other) may be involved. Refer to the sample equipment evaluation worksheet at the end of this section.

3. Sequence of Lockout/Tagout Procedure

Notify all *affected employees* (staff working near and around work to be done) that a lockout system is going to be utilized and the reason thereof. The *authorized employee* (staff qualified to do the work) shall know the type and magnitude of energy that the machine or equipment utilizes and shall understand the hazards thereof.

If the machine or equipment is operating, shut it down by the normal stopping procedure. Operate the switch, valve, or other energy-isolating devices so that the equipment is isolated from its energy source. Stored energy (springs, elevated machine members, rotating flywheels, hydraulic systems, and air, gas steam, or water pressure) must be dissipated or restrained by a method such as repositioning, blocking, bleeding down, etc.

Lock and tag out the energy-isolating devices with assigned individual locks and tags.

To ensure that all energy sources have been deactivated, ensure that employees are not exposed, and then operate the push button or other normal operating controls to make certain the equipment will not operate. **Caution:** Return operating controls to neutral or off position after the test.

The equipment is now locked and tagged out. Repairs or maintenance work may begin.

After the servicing and/or maintenance is complete and equipment is ready for normal produc-tion operations, check the area around the machine to ensure that no one is exposed.

After all tools have been removed from the machine or equipment, guards have been reinstalled, and employees are in the clear, remove all lockout and tagout devices to restore energy to the ma-chine or equipment.

Notify the affected employees that work has been completed.

4. Procedure Involving More Than One Person

In the preceding steps, if more than one individual is required to lock out or tag out equipment, each shall place his or her own personal lockout device on the energy-isolating device(s). When an energy-isolating device cannot accept multiple locks or tags, a multiple lockout or tagout device (hasp) may be used. When lockout is used, a single lock may be used to lock out the machine or equipment with the key being placed in a lockout box or cabinet that allows the use of multiple locks to secure it. Each employee will then use his or her own lock to secure the box or cabinet. As each person no longer needs to maintain his or her lockout protection, that person will remove his or her lockout protection and will remove his or her lock from the box or cabinet.

5. Personnel Authorized to Lock out/Tag out

_____

Name/Title

_____

Name/Title

_____

Name/Title

_____

Name/Title

6. Training

Appropriate employees shall be instructed in the safety significance of the Lockout/Tagout procedure [Names/Job Titles of Employees Authorized to lock out or tag out]. Each new or transferred affected employee and other employees whose work operations are or may be in the area shall be instructed in the purpose and use of the Lockout/Tagout procedure.

All authorized and affected employees shall be trained annually in the correct implementation of this program and its elements. This must include:

- The types of energy control procedures utilized and when the control procedure is being implemented.
- A review of the purpose of the procedure and the importance of not attempting to start up or use the equipment that has been locked or tagged out.

The training for authorized employees must include:

- Details about the type and magnitude of the hazardous energy sources present in the workplace.
- The methods and means necessary to isolate and control those energy sources.

A "certification" will be prepared with the names and dates of training. Refer to the sample training rosters, one designed for group training, and one for individual training to document occurrence, at the end of this section.

7. Periodic Inspection

At least annually, there will be an inspection and verification of these procedures:

- Any deviations from established procedures will be noted and rectified.
- A certification will be prepared with the date of each periodic check documenting findings and remedial actions.
- These records should be kept on hand for at least the current and previous inspection.

8. Contractors

Contractors must have their own Lockout/Tagout policy. No electrical or mechanical (HVAC, elevator, etc.) contractor will be employed or allowed to begin work until it has shown that it has a working Lockout/Tagout program in place that satisfies all the requirements of Title 29 CFR 1910.147. Facility employees on the premise will be instructed to comply with the contractor's Lockout/Tagout policy. The contractor's program shall include the written procedures; the appropriate training; the implementation of said program; and yearly inspection and certification.

SAMPLE FORM: EQUIPMENT EVALUATION FOR LOCKOUT/TAGOUT PROCEDURES

Date: _____ Name of person conducting evaluation: _____
Equipment and location:
_____
_____

Location and type of energy isolating device:
_____
_____

Can the equipment be locked out, if so, how? _____
_____

If the equipment cannot be locked out, why not? _____
_____

Can it be tagged out? _____
Are there any special precautions that must be followed? _____
_____
_____

Notes: _____
_____
_____

SAMPLE FORM: TRAINING ROSTERS FOR LOCKOUT/TAGOUT PROCEDURES

Group Training Roster [See In-Service in Section E]
Date of Training:_____/_____/_____Name of Trainer:_____

Name of Employee	Title	Signature
Affected Employees Trained		
_____	_____	_____
_____	_____	_____
_____	_____	_____
_____	_____	_____
_____	_____	_____
_____	_____	_____
_____	_____	_____
_____	_____	_____
_____	_____	_____
_____	_____	_____
_____	_____	_____
_____	_____	_____
_____	_____	_____
_____	_____	_____

Authorized Employees Trained

_____	_____	_____
_____	_____	_____
_____	_____	_____
_____	_____	_____
_____	_____	_____
_____	_____	_____
_____	_____	_____
_____	_____	_____
_____	_____	_____
_____	_____	_____
_____	_____	_____
_____	_____	_____
_____	_____	_____

Individual Training Roster [See In-Service in Chapter 5]

Date of training: _____ /_____ /_____   Name of trainer: _____

Name of employee: _____

Title of employee: _____

Employee signature: _____

Trained as affected or authorized employee (circle one)? Affected/authorized

SAMPLE FORM: LOCKOUT/TAGOUT PROCEDURES ANNUAL INSPECTION

Date of inspection: _____/_____/_____ Inspected by: _____ (must be owner or plant manager)

Procedures Found in Need of Enforcement or Improvement	**Remedial Action Taken**
_____	_____
_____	_____
_____	_____
_____	_____
_____	_____
_____	_____
_____	

Signature of Inspector

_____

XIII. MACHINE GUARDING

A. INTRODUCTION

OSHA inspectors spend a great deal of their extensive safety training on machine guarding of every kind. There are a number of guarding standards. They range from very equipment specific, to general rules of thumb. Machine guarding violations have graced the top ten most frequently cited standards list in all industries for the past decade. Machine guarding violations may not be the most relevant or serious safety problem in a facility, or may not affect workers' compensation premiums. However, inspectors will seek them out and obsess over any violation that can possibly be found, usually confounding the facility operator's sense of reason.

B. DISCUSSION

Some typical examples of potential violations include:

- The old out-of-service bench grinder in the back of the maintenance shop or in the pile in the garage along with the deceased residents' belongings (it should have guards on both sides, and there shall be no more than one-eighth inch space between the abrasive wheel and the tool rest; if not, lock it out and label it "out of service").
- The resident fan cage that is held together by a bobby pin since it separated.
- The gap in the manufacturer's guard on the emergency generator.
- The refrigeration compressor guards that have been keeled wider or shifted over time.
- The rotating exposed parts behind the dryers after repairs or maintenance were completed but the covers were not reinstalled.
- The Hobart mixer that did not come with a guard (call Hobart, 718-545-2240, with model number, material list number, and specification number for quote, cost ranges from $1,500 and up, but it is cheaper than a fine, and precludes the unlikely but highly costly potential of a serious workers' compensation claim with third–party potential).
- Antique resident fans that have inadequate cages (wires no more than one-half inch apart).
- The guard on compressors and wall and window exhaust fans.
- The guard on HVAC and roof-top evaporation units.
- Other countless old or new manufactured design unaltered equipment that may have inadequate guarding per OSHA standards (can crushers, garbage disposal).
- The guard on the meat slicer.
- Any and all belts, pulleys, or moving parts exposed from ground level up to 7 feet in height without appropriate guarding or enclosure.

This fixation with irrelevant issues may very well seem aberrant to long-term care operators and their staff. Keep in mind, however, that the relevant issues in health care are often not standardized, but are politically charged and controversial (ergonomics, workplace violence prevention, tuberculosis). And although there is a Bloodborne Pathogens Standard, it is still relatively new and applied in a wholly unfamiliar environment to the inspector. Meanwhile, old faithful standards that have withstood the test of time, however arcane, are safe and familiar ground for the inspector. Thus, inspectors resort to those standards that had been commonly applied in manufacturing settings.

There is one semivalid argument in defense of this nitpicking strategy. Inspectors are unable to cite facilities for deficient safety management, i.e., lack of leadership, direction, accountability, responsibility, resources, budget, personnel, etc. In the absence of the ability truly to identify (a sublime task) and cite the root causes inherent in the problems of an organization, inspectors resort to existing standards that have *stuck*. This is the trade-off at the basis of the OSHA PEP (Program Evaluation Profile) new inspection strategy (see Chapter 2, Section III).

C. SUMMARY

It would behoove facility operators to conduct a comprehensive wall-to-wall machine guarding survey and instill in staff, especially maintenance personnel, the strict importance of viewing guarding of all moving mechanical parts with great conservatism. As a general rule, all equipment under 7 feet must be guarded in such a way to prevent any body appendage from entering the "line-of-fire" (moving parts). Fan cage wires must not be more than one-half inch wide. Other types of guard spacing requirements or formulas are regulated by how far into the cage or how wide or narrow the space between guards is from the moving parts of the machine. (The farther away, the larger the spacing allowed, and vice versa.) The closer to the line of fire, the smaller the gap will be required. Without indulging in technical details, the rule of thumb is to prevent the body appendage from entering the line of fire.

XIV.   PERSONAL PROTECTIVE EQUIPMENT

A.   GOAL

Employers must have a comprehensive program to ensure that the appropriate PPE is selected, made available to employees, maintained, and that proper PPE use training has been provided to employees who must use them.

B.   SUMMARY

The PPE standard was revised in April 1994 to formalize the steps necessary in selecting, maintaining, providing, and training employees on the proper use of PPE. Inspectors have begun to review the written PPE program to reflect the strengthened requirements of this revised standard. Inspectors continue to observe and verify that the proper PPE is selected, sized, available, and being used properly.

Although employees are to be trained in the proper use of PPE, inspectors do not generally seek PPE training documentation. Inquiries on PPE are spurred by employee interviews that may point to deficiencies in PPE training. PPE training is also specified in particular standards, i.e., Bloodborne for universal precautions, Hazard Communication for protection against chemical exposures, Respiratory Protection, Lockout/Tagout, etc. whereupon specific class content, frequency, and attendance criteria must be followed and well documented. (See In-Service Records, Chapter 5 in this manual.)

C.   UPDATE

The standard is being revised to specify explicitly that the employer is responsible for the cost of the equipment. It has never been regulated as such.

The following guide will help generate documentation that parallels PPE requirements under more hazard specific standards (Bloodborne, Hazard Communication, Tuberculosis, etc.). It also provides a comprehensive review of PPE needs throughout the operation of the facility. Long-term care facilities have not, as yet, been cited for a lack of a comprehensive PPE program. However, it is on the books, and it is very applicable.

D.   CONTENTS

A guide to personal protective equipment compliance is presented in Section XIV.E. This guide assists users in developing a document trail that would serve as very strong evidence of a formalized and compliant personal protective equipment program. At the end of the guide are sample certifications for workplace hazard assessment and training for PPE.

E.   GUIDE: PERSONAL PROTECTIVE EQUIPMENT

Introduction

This guide will assist you in developing and implementing an OSHA-compliant Personal Protective Equipment Program in accordance with the revised (April 6, 1994) Title 29 CFR 1910.132 to 1910.140.

The following includes a step-by-step assessment with worksheets. The assessment is mandatory for each type of protective equipment used in the workplace. Also included is a menu of components that make up your Personal Protective Equipment Program.

Data and Documentation Review

Review your OSHA 200 Log, workers' compensation claims, and accident/incident reports to determine the types of injuries your facility has experienced. Look for trends, common injuries, non-routine injuries, and even potential injuries that may not have been manifested yet. (Refer to Chart 1: Injury Data Review Sheet.)

Review previous industrial hygiene surveys, safety evaluations, or safety committee meeting notes to provide as much of a background as possible for the on-site evaluation.

On-Site Hazard Assessment

Audit each work location and job task for safety and health hazards. Assign an individual to observe work processes and identify potential hazards for each job task. The observations should be task and area based. Identify the individual, as well as any person working in or passing through the area who are similarly directly exposed. The observer should be acquainted with the process and should ask

---

CHART 1
Injury Data Review Sheet

OSHA 200 Log Review

Top 10 Most Frequently Reported Injuries

1. _____
2. _____
3. _____
4. _____
5. _____
6. _____
7. _____
8. _____
9. _____
10. _____

OSHA 200 Log Review

Top 10 Injuries: Time Lost

1. _____
2. _____
3. _____
4. _____
5. _____
6. _____
7. _____
8. _____
9. _____
10. _____

Workers' Compensation Most Frequent Injuries: Most Costly Injuries

1. _____
2. _____
3. _____
4. _____
5. _____
6. _____
7. _____
8. _____
9. _____
10. _____

1. _____
2. _____
3. _____
4. _____
5. _____
6. _____
7. _____
8. _____
9. _____
10. _____

Previous Review Notes from Surveys, Committee Meetings, etc.:

_____
_____
_____
_____
_____

questions of those being observed on how they should conduct their tasks, and what health and safety concerns they may have. Any relevant comments should also be noted.

Inventory and note all sources of:

- Motion
- Temperature extremes
- Chemical exposure
- Falling objects
- Sharp objects
- Rolling or pinching objects
- Ionizing and nonionizing radiation

Inventory and note all sources that result in:

- Impact
- Penetration
- Compression
- Chemical and harmful dust exposure

Refer to Chart 2: On-Site Hazard Assessment.

Elimination of Identified Hazards or Potential Risks

Attempt to eliminate the need for protective equipment by installing engineering controls, or re-designing workstations, or administratively rotating and rescheduling workers to spread marginal exposures so that ultimate risks are negligible. *Consider:*

- Machine guards
- Netting
- Ventilation systems
- Chemical substitution
- Work task redesign

Record findings in the hazard elimination portion of the hazard assessment chart at the end of this section.

Selection of Appropriate Personal Protective Equipment

If the hazard cannot be adequately engineered out of the process, or if PPE is used in conjunction with existing hazard abatement efforts, select the appropriate PPE, i.e., eye, face, foot protection, protective clothing, etc.

If you determine that a type of PPE is required to protect against a chemical hazard, you must determine the chemical resistance of the PPE material or article to the specific chemical. You will need to review the chemical permeation rates and breakthrough times for the protective equipment. *Breakthrough time is the time it takes a chemical to pass through protective equipment.* Permeation rate is the speed by which a chemical moves through a sample of protective equipment.

If you are unsure of the specific type or level of PPE you should use, refer to the MSDS for guidance. There is a section on PPE on the MSDS that should instruct you regarding what type of PPE is necessary, and when. If you are unsure of the breakthrough times and how to determine chemical resistance of the PPE, or if you are still uncertain of the type and level of PPE, or if the MSDS is not clear, or if the situation is more complex than the MSDS addresses, you may also contact the author's office and speak with either one of our Health and Safety Advisors, Ralph Ciano or Elsie Tai for more specific guidance.

CHART 2
On-Site Hazard Assessment

Date: _____   Observer: _____   Duration: _____

Time: _____   Location: _____   Task: _____

Source of Hazard and Description (check and make notes):

_____ Motion _____

_____ Temperature Extremes _____

_____ Chemical Exposure _____

_____ Falling Objects _____

_____ Sharp Objects _____

_____ Rolling or Pinching Objects _____

_____ Ionizing/Non-ionizing Radiation _____

Hazards Resulting in:

Impact

_____

Penetration

_____

Compression

_____

Chemical and Harmful Dust Exposure

_____

_____

_____

_____

_____

Type of Injury	Location of Injury	Seriousness of Injury
_____ Cuts/laceration	_____ Hand/finger	_____ Amputation
_____ Abrasions	_____ Face	_____ Fracture
_____ Punctures	_____ Eye	_____ Blindness
_____ Skin absorption	_____ Head	_____ Scarring
_____ Chemical burn	_____ Skin	_____ Dermatitis
_____ Thermal burn	_____ Foot	_____ Other-_____
_____ Temperature extremes	_____ Arms/legs	_____ Other-_____

Can Equipment or Task Be Modified to Design the Hazard out of the Process?   Y/N
(machine guards, netting, ventilation systems, chemical substitution, task redesign)

_____

_____

_____

Type of PPE Required:

_____

_____

_____

_____

Please note that required PPE is the financial responsibility of the employer, and the cost cannot be passed onto employees unless equipment is deemed nonessential. Record selected PPE in the appropriate portion of the Hazard Assessment Chart.

Note: The use of respirators or hearing protectors to reduce or eliminate chemical exposure or noise hazards requires additional documentation, medical clearance and baseline audiograms, and implementation of a Respiratory Protection Program, or a Hearing Conservation Program, respectively. These elements are not included here since they are more client specific and require a considerable amount of space.

If you believe that you may require the use of respirators or hearing protection, please contact the author's office and speak with Ralph Ciano or Elsie Tai. We will be able to assist you in determining your need, furnish you with the documentation and guidance you will require to develop, implement, and maintain such mandatory programs, and provide you with further references as necessary.

Employee Training

All affected employees must be trained annually on the need for and the appropriate use and maintenance of the PPE issued to them.

SAMPLE FORM: CERTIFICATION OF WORKPLACE HAZARD ASSESSMENT

As per OSHA Standard 1910.132, Personal Protective Equipment Standard

I hereby verify that I performed the required workplace hazard assessment as documented in the following pages identified as [Chart 2: On-Site Hazard Assessment] at:

_____, located at _____

_____, on _____, 19_____.

     Name: _____

     Title:_____

     Date:_____

SAMPLE FORM: CERTIFICATION OF TRAINING FOR PERSONAL PROTECTIVE EQUIPMENT

[See Chapter 5, In-Service]

TRAINING CONTENT:

- When PPE is necessary
- What PPE is necessary
- How to don, doff, adjust, and wear PPE properly
- The limitations of the PPE
- The proper care, maintenance, useful life, and disposal of the PPE

Facility Name: _____

Training Date: _____   Trainer: _____

Employee Name	Title	Department
_____	_____	_____
_____	_____	_____
_____	_____	_____

_____   _____   _____
_____   _____   _____
_____   _____   _____
_____   _____   _____
_____   _____   _____
_____   _____   _____
_____   _____   _____
_____   _____   _____
_____   _____   _____
_____   _____   _____

4 General Duty Nonstandard Issues

I. INTRODUCTION

A. GOAL

The General Duty Clause allows OSHA inspectors to cite hazards that are serious but that are not specifically regulated by a standard. To be cited, the hazard must be a *recognizable hazard*, likely to cause death or serious physical harm. There also must be abatement methods *available* that the employer has not employed.

B. SUMMARY

The General Duty Clause Section 5(a)(2) states that "Each employer shall furnish to each of his employees employment and a place of employment which are free from recognized hazards that are causing or are likely to cause death or serious physical harm to his employees." General Duty Clause citations are only to be issued when there is no standard that applies to the particular hazard. *The hazard, not the absence of a particular means of abatement*, is the basis for a General Duty Clause citation. All applicable abatement methods identified as correcting the same hazard shall be issued under a single General Duty Clause citation.

The key words are *hazard, reasonably recognizable, unabated, likely*, and *serious*. There must be a hazard, that is recognizable, by a reasonable person, that is left unabated, and that is likely to cause serious injury. Again, General Duty Clause citations are only applicable when no other standard has jurisdiction over the issue.

Inspectors tend to shy away from General Duty Clause citations for several reasons. First, the citations are easier to contest than citations of violations of standards since no standard exists decreeing what shall and shall not be done specifically. General Duty Clause citations are more likely to be arguable, in terms of if the hazard exists, if that hazard is serious, if a reasonable person would have recognized the hazard, and if all those elements are true and present, if it can, in fact, be abated. These are *exotic* citations.

General Duty Clause citations pose challenges to inspectors and their supervisors who do not appreciate the additional work and controversy such citations are apt to generate. Thus, inspectors and their home offices tend to ignore relevant violations that have not been regulated by standards. Instead, they prefer to cite odd and less relevant hazards that have been regulated by a specific standard and have a strong citation history, which precedent has established, thereby eliminating the potential for controversy and the work necessary in dealing with something new or different.

C. UPDATE

Ergonomics, tuberculosis (TB), and workplace violence prevention are relevant issues that will be reviewed in the course of a routine OSHA inspection of a long-term care facility. Since there are no standards for these issues, any citations must be based on the General Duty Clause. Citations are highly unlikely for the various reasons discussed above. However, citations had been given to Beverly Enterprises for ergonomic violations, and to a number of hospitals for TB violations.

Workplace violence prevention citations have not been issued, as of yet. However, OSHA closely follows developments in these subjects and issues guidelines on them.

Depending on the ultimate inspection strategy protocol adopted, the home office agency culture, and the individual inspector, it is more likely that deficiencies in these areas would not lead to General Duty Clause citations per se, but more numerous and/or severe citations in other findings, which have established standards.

This was the trade-off seen in the OSHA PEP (Program Evaluation Profile). This is the new inspection strategy pilot program. The PEP, which is discussed and can be found in Chapter 2, Section III of this manual, is a comprehensive health and safety program assessment tool. It evaluates the quality of the health and safety management system, whose components are not mandated, i.e., safety program, safety policy, safety committee, adequate budgeting, personnel, resources, etc. The tool produces a score from 1 to 5 with 5 the highest. The idea was that if the facility scored a 3 or better, the inspector would then only conduct a "focused inspection" on the issues that were flagged (i.e., TB conversions or needle sticks or exposure incidents or combative resident incidents on the OSHA 200 Log). This minimizes the scope of the inspection. However, if the facility scored less than a 3, then the inspector would proceed with the traditional wall-to-wall inspection looking for any and all valid citations, regardless of their operational relevance.

It is unclear to what extent inspectors now use the PEP, or revert to the traditional inspection methods, although they began using it in New York in the spring of 1997. The return to the Interim Plan does not bode well for long-term care facilities' probability of an inspection. Nor does it enhance the trend toward more cooperative and experimental programs that OSHA has been embarking on since the Senate hearings on OSHA Reform and the congressional mandate for reform.

Nonetheless, these General Duty Clause issues, ergonomics, TB, and workplace violence prevention, should be managed not only for the sake of OSHA compliance, but because these are the leading employee health and safety issues for the long-term care industry. Ergonomics and workplace violence (in terms of combative residents in addition to other types of violence) are most often the leading causes of workers' compensation claims in long-term care, along with slips, trips, and falls. Although the risk of TB is very low in most facilities, the potential does exist and the severity can be very serious, especially if it involves multidrug-resistant strains, not to mention the employee and public hysteria that must be managed and contained should the *mere thought* of TB be introduced into a facility population.

The following sections provide a comprehensive review and discussion of these issues, together with supplemental policy guides to managing all three nonstandardized but highly relevant issues.

II. ERGONOMICS

A. GOAL

Back injury is the leading cause of workplace injuries, loss time injuries, and contributes to the majority of workers' compensation claims and costs, particularly in long-term care. In an effort to reduce the high rate of worker back injuries, and related muscle sprains and strain injuries, as a result of repetitive lifting and improper patient handling, employers are required to take some measures in an attempt to address the problem.

B. SUMMARY

There is no ergonomic standard per se. The controversial draft Ergonomic Standard was published November 1999. Public hearings were held with short notice in the early spring of 2000. The controversy surrounding this standard is severalfold. Opponents say that there is no legitimate and credible causal relation evidence available proving the direct correlation between cumulative trauma disorders and the type of work conducted. Others who concede that there is a relationship, object to the extensiveness of the proposed OSHA rule in mandating employee benefits (90% wage replacement and 6-month job preservation). Meanwhile, others criticize that the triggers and definitions throughout the proposed rule are too vague and/or arbitrary. There also seems to be an urgency in the proceedings in an effort to get this rule completed within the period of the current administration. There are complaints about insufficient notice to review cost benefit analysis and the rule itself, and to provide appropriate responses, including public hearings.

Thus, in a nonspecific way, but in response to the General Duty Clause and the fact that back injuries are the leading cause of occupational injuries in the industry, nursing homes are required to have some sort of ergonomic program. It could be in the form of a written ergonomic program, ergonomic analysis in safety committee, proper lifting techniques training, purchase, use, and maintenance of assistive lift devices, policy guidance on how certain residents are to be transferred, etc.

C. UPDATE

The controversial draft Ergonomic Standard has not affected the likelihood of a facility getting a General Duty Clause citation for ergonomic deficiencies. Until there is a published standard in place, the probability and nature of citations have not changed. There is heightened awareness on the part of inspectors and operators. Nursing homes can be cited for "repetitive lifting hazards." More likely, however, in the present political and OSHA inspection strategy climate, poor ergonomic management will most likely not result in ergonomic citations, but in more citations in standard areas. It is best to have a comprehensive ergonomic policy that covers the above-mentioned elements in a cohesive program that monitors trends, investigates incidents, provides abatement possibilities, and follows up on effectiveness. Also available from the author's office is the "OSHA Boston Regional Office Instruction to Inspectors," which provides guidance and procedures for ergonomic inspections. This has not yet been adopted by federal OSHA.

D. CONTENTS

An ergonomics program to address patient handling injuries in nursing homes, as distilled from OSHA Region II Technical Division (substantively from OSHA-generated guidelines that address only patient handling) is presented in Section II.E.

A sample comprehensive ergonomics policy for long-term care, Section II.C is designed to be a comprehensive ergonomic policy for a facility to include all staff and departments, but which embodies the OSHA Region II Technical Division Guide to Address Patient Handling Injuries in Nursing Homes, or summarized in the above document.

E. ERGONOMICS PROGRAM TO ADDRESS PATIENT HANDLING INJURIES IN NURSING HOMES*

1. Management Commitment and Employee Involvement

1. Management demonstrates involvement by placing a *priority* on eliminating ergonomic hazards.
2. Management commitment is demonstrated by assigning and communicating *Responsibility* for the ergonomic program to all managers, supervisors, and employees. *Accountability* mechanisms are in place.
3. An effective written program for job safety, health, and ergonomics is in place, outlining the nursing home's goals, which is communicated to all employees. All managers,

---

*Substantively from OSHA Region II Technical Division Guidelines.

supervisors, and employees involved *know what is expected of them.* This program should include policies and procedures about:

a. Orientation of employees that includes education in injury prevention.
b. Continuing injury prevention education.
c. Methods of transfer/lift to be used by all staff.
d. Modified (light-duty) work, and postinjury return to work programs.
e. Compliance with transfer and lift procedures.
f. Procedures for reporting of early signs and symptoms of back pain, and other musculoskeletal injuries.

This written program should be reviewed and updated at least annually and whenever necessary to reflect new or modified tasks and procedures that affect worker exposure to ergonomic hazards.

4. Employee involvement in the ergonomic program is encouraged through a complaint/suggestion program, prompt and accurate reporting of injuries, establishment of safety and health committees, and training of employees in the skills necessary to analyze jobs for ergonomic stress.

5. Procedures are implemented to evaluate regularly the effectiveness of the ergonomics program, and to monitor success in meeting goals and objectives.

This process includes review of recorded OSHA 200 injuries, workers' compensation and insurance reports, and reports from employees of unsafe working conditions. Also included are regular (monthly) walk-around inspections, safety and health committee meetings, and employee surveys regarding worksite changes.

2. Work Site Analysis

Work site analysis identifies existing hazards and conditions, work habits that create hazards, and areas where hazards may develop. Also included is close scrutiny and tracking of injury and illness records to identify patterns of trauma or strains.

1. Nursing home gathers relevant information on ergonomic solutions to patient handling problems.
2. Baseline screening surveys using a checklist are conducted to evaluate ergonomic risk factors and determine which tasks are most stressful and need improvement.
3. Job analysis is performed by persons skilled in evaluating ergonomic risk factors associated with patient handling requirements. Evaluation is conducted during peak lifting and transferring times.
4. Changes are implemented to avoid the most stressful patient transfers.
5. Periodic surveys and follow-ups are conducted to evaluate changes.
6. Screening surveys/checklists are utilized to evaluate workplace stressors.
7. After each patient handling injury, a determination is made as to whether a task can be modified to reduce future risks.

3. Hazard Prevention and Control

Ergonomic hazards are prevented primarily by proper selection of equipment, effective use of assistive devices, and by implementation of proper work practices. The equipment must be available in sufficient quantities, convenient for use, and properly maintained.

1. Patient chairs that allow the nurse to place feet under chair; chair back is low enough to permit lifting access.
2. Mobile chairs, wheelchairs, and commodes have functional brakes, mobile arms, and footrests that do not obstruct movement.

3. Transfer surfaces are approximately at the same level, for example, toilet seats are raised so that the heights of the wheelchair and toilet seat are the same. Bath stretchers are height adjustable to allow horizontal transfers.
4. Hoists are selected based on nurse and resident evaluation of ease of use, comfort, and safety. Hoists are routinely maintained, are conveniently located and available, with suitable lifting attachments and/or slings, such as:
   a. Fixed hoists, often used in bathrooms due to limited space;
   b. Hoists on a track, which can also be operated by patients;
   c. Mobile hoists.
5. An adequate number of hoists is available for the transfer of residents who are dependent or can only provide minimal assistance in the transfer.
   a. At least one portable resident hoist with a maximum capacity is available for extremely heavy patients. A backup hoist is available when the primary lift is out of service.
   b. One hoist, with an adequate number of slings, is available in each wing and easily accessible.
   c. At least one of the portable hoists in the nursing home should have the functional capacity to lift a resident from the floor.
   d. A digital scale is attached to one mechanical hoist to allow the weighing of dependent residents during routine transfers.
6. Guidelines have been developed to instruct nurses in which type of hoist or sling would be appropriate for particular patients.
7. Nursing supervisors are trained so that they understand the importance of using patient handling devices and the techniques for using them.
8. "Sliding boards" are used to ease transfers from wheelchair to bed and bed to wheelchair when the resident has the cognitive and upper body strength to assist.
9. Transfer belts are used where appropriate.
10. Patient transfers are minimized. Shower chairs are utilized, which are also compatible with toileting facilities. Weighing stations allow residents to remain in their chairs while being weighed on a wheelchair ramp.
11. Good ergonomic bed design is evident through convenient location of controls, adjustable bed heights to wheelchair, and sufficient foot clearance.

4. Work Practices/Administrative Controls

1. Supervisors are familiar with patient handling guidelines and enforce company rules.
2. Injuries are accurately recorded on the OSHA 101.
3. Care plans are specific in assessing patient handling requirements, and are communicated to affected employees prior to patient handling. A practical system for communicating changing assessment results is developed, and special attention is given for residents who have special handling needs, such as residents who have weakness on one side from a stroke.
4. Patient clothing is selected that aids transfers (nonslippery; clothing handles for the unconscious residents; adaptive clothing for ease of changing; absorbent pads for incontinent patients).
5. Proper patient transfer techniques are utilized (patient is properly positioned prior to transfer; patient handling slings or transfer belts are utilized; patient clothing is never adjusted while transferring).
6. Slips and trips are minimized by eliminating uneven floor surfaces where possible, using large wheels on transport equipment, creating nonslip surfaces in toilet/shower areas, and enforcing a policy of immediately cleaning up fluids spilled on a floor.
7. A policy exists that ensures that patient handling help is provided to a nurse any time it is requested.

8. Restricted "light-duty employees" are not involved in transferring patients.
9. An effective program for facility and equipment maintenance is implemented that minimizes ergonomic hazards and includes:
   a. A preventative maintenance program for patient handling devices.
   b. Maintenance whenever employees report problems; sufficient spares exist for out-of-service equipment.
   c. Implementation of housekeeping programs that minimize slippery work and slip/falls.

5. Medical Management

Proper medical management is necessary to eliminate or reduce the risk of patient handling injuries through early identification and treatment, and to prevent future problems through rehabilitation and training. The program is supervised by a person trained in the prevention of cumulative trauma disorders, and is a consultant to the safety and health committee. The program is periodically evaluated and includes:

1. Accurate injury and illness recording.
2. Early recognition and reporting.
3. Conservative treatment with specific transferring restrictions during recovery.
4. For disabling injuries, early (e.g., 2 days) referral to specialist in physical medicine and rehabilitation.
5. Systematic monitoring, return to work only after transferring skills have been reassessed.
6. Baseline health assessment for comparing changes in health status; replacement evaluation of patient handling skills.
7. Medical management participation in the development of a list of light-duty jobs, to assist management in assigning light or restricted duty jobs.

6. Training and Education

The object is to ensure that managers, supervisors, and health care providers are sufficiently informed about the ergonomic hazards to which they may be exposed and thus are able to participate actively in their own protection. Training programs should be designed and implemented by qualified persons. The program should include an overview of the potential risk of back and other musculoskeletal injuries, their causes, and early symptoms, and means of prevention and treatment. Appropriate "train the trainer" instruction is required for personnel responsible for providing training. Training should be presented in a language, and at a level of understanding, appropriate for the individuals being trained, and should include opportunities for interactive questions and answers with the trainer.

1. The orientation/new employee program includes job site evaluation of transferring technique by a person skilled in the art and science of transferring patients. Feedback is provided. Basic training in handling patients is given. Supervising nurses are trained to develop polices and procedures for their units and to train their nursing staff:
   a. Never transfer patients when off balance.
   b. Avoid heavy work with spine rotated.
   c. Lift loads close to the body.
   d. Avoid vertical "dead-lifts."
   e. Never risk overexertion with a patient who is resistant; use assistance.
   f. Use team lifts and mechanical devices where necessary (patients over 150 lb are always considered "heavy").
   g. Properly place and adjust equipment used to handle patients.

    h. Always bring patient toward you, never away.

    i. Do not lift fallen patients alone; get help, consider mechanical assistance.

2. Employees (nurse's assistant, LPN, RN, etc.) are also provided:

    a. Proper training for assisting walking patients (stand on weaker side, close to patient, take load on hip rather than on back).

    b. Training in appropriate transfers and lift techniques in confined spaces, such as shower stalls and toilet areas.

    c. Training in when and how to use mechanical hoists (e.g., for patients over 150 lb who cannot support their own weight).

    d. Initial and periodic (e.g., monthly) programs on back care and transferring techniques.

    e. Training on the appropriate patient handling devices to use with particular patients and the proper use of this equipment.

3. Training in emergency patient handling is given, such as for patients who have fallen, have spasms, are combative, or exhibit any unpredictable behavior.

4. Training and feedback are provided on use of patient handling devices.

5. Annual refresher training is provided and addresses specific needs. Provisions are made to train absent employees.

6. Staff physical fitness is encouraged.

7. Training on early identification, reporting, and conservative treatment of cumulative trauma disorders is provided to all workers, supervisors, and management on an annual basis.

F. SAMPLE POLICY GUIDE: COMPREHENSIVE ERGONOMIC POLICY FOR LONG-TERM CARE FACILITIES

About This Guide

This guide was developed to assist long-term care facilities in designing an overall ergonomic program for the *entire* facility. This guide embodies the elements of the OSHA Region II Technical Division "Ergonomic Program to Address Patient Handling Injuries in Nursing Homes," mostly under the Nursing Department section in the guide as that document only addresses nursing. This sample policy guide is written in conjunction with the sample Health and Safety Policy from Chapter 2, Section IV of this manual. *It is best to correlate an ergonomic program within the structure and foundation established by an existing safety program.*

How to Use This Guide

The components that are fundamental to an ergonomic program are marked [basic]. Ensure that these [basic] aspects are included in designing a facility's policy. The [optional] designation denotes elements that should be considered to enhance the comprehensive nature of the program.

Introduction

Soft tissue injuries are the leading cause and cost of work-related injuries in long-term care. The controversial Ergonomic Standard being drafted by OSHA will probably not become final for quite some time. In the interim, OSHA inspections of long-term care facilities concentrate on ergonomics, among other more traditional and less relevant subjects. It is unlikely to receive citations for ergonomic deficiencies. Inspectors prefer more typical, traditional, nondisputable citations, like machine guarding or chemical labeling. Nonetheless, **Ergonomic Programs will be reviewed during an OSHA inspection. It is also the key to the practical management of those types of injuries and containing their substantial costs. A well-documented ergonomic program will also serve to comply with Joint Commission Standards PI.3.3.2: Risk Management Data in Improving Organization Performance, and EC.1.3 and EC.2.2: Safety Management.**

Mission Statement [basic]

The management of_____Nursing Home is committed to providing excellence in long-term care. Optimal ergonomics is an integral part of overall operational performance. The management strives to provide effective leadership in maintaining optimal ergonomic conditions for our staff, as it does in maintaining a safe and healthful home and living conditions for our residents. Optimal ergonomics and quality are organization priorities. Refer to our Health and Safety Policy for an overview of all the elements in the health and safety management system as they apply to this Ergonomic Policy.

Ergonomic Responsibilities and Accountability [basic]

Supervisors and managers are responsible for providing adequate job-specific training, counseling, positive feedback, and supervision to employees to *ensure that the proper protocols and equipment are consistently followed and used.* Staff are expected to follow established work procedures and equipment designed to optimize ergonomics. All staff and management are responsible to the residents, each other, and themselves for remaining conscious of, practicing, and ensuring good ergonomic practices.

Employees are encouraged to relay concerns and suggestions on ergonomic hazards or improvements. Supervisors and managers must respond to these concerns in a positive manner. Supervisors shall ensure that the message is channeled to the Safety Committee, if necessary. Supervisors and/or the Safety Committee shall explore the potential of the hazard or viability of the suggestion, and take measures to implement the change, or explain to the employee initiating the concern or suggestion why it is not implemented.

[optional] Ergonomic performance is part of the safety performance section of the overall annual evaluation. Ergonomic performance is expected and evaluated like other performance criteria on the annual evaluation, i.e., conduct, hygiene, attendance, delivery of care, etc. and the consistency of these activities.

Ergonomic Committee/Safety Committee [basic]

There is an Ergonomic Committee chaired by [name] _____, [title] _____ . It is responsible for the elements of this Ergonomic Program, and reports its findings and activities to the facility Safety Committee. It may require approval from the Safety Committee to embark on certain abatement or special assessment activities. It meets at a frequency of _____ times per _____. The Ergonomic Committee is under the direction of the general Safety Committee. The Safety Committee reports to the Chief Executive on a regular basis. Good communication between these two groups is essential to the success of the overall Health and Safety Program.

[or]

For facilities that are smaller and/or less complex, these responsibilities can be assumed directly by the Safety Committee, itself, chaired by [name] _____, [title] _____. The Safety Committee implements this Ergonomic Program instead of delegating to an Ergonomic Subcommittee. It meets at a frequency of _____ times per _____.

The Ergonomic Committee is responsible for implementing the elements of this policy, including assessing ergonomic hazards through data review, worksite analysis, equipment review, and employee feedback; developing abatement methods and implementing them; determining abatement effectiveness; responding to employee complaints or inquiries; and monitoring the effectiveness of the overall program.

Data Review [basic]

A review of the OSHA 200 Logs and accident/incident records for the past several years is conducted periodically to quantify and identify trends, frequency, severity, and potential common causes of incidents. This serves to benchmark statistical performance over time and, more importantly, to help prioritize intervention activities based on the existing hazards that appear to contribute to the most frequency, and severity.

Note: The quality of this data review is wholly dependent on the quality of the data collection and accident investigation methods. Incomplete and inaccurate data will undermine this review process. Poor root-cause analysis during cursory accident investigations will not bear fruit to effectively identify real systemic causes of incidents. Ensure that data collection, including accident investigations, is thorough, accurate, and, in the case of root causes, truly correct and sufficiently substantive.

Workplace Analysis or Hazard Assessment [basic]

Inspection [basic]

Conduct a comprehensive ergonomic inspection of all work areas, and work procedures, including storage areas, stocking tasks, receiving, offices, kitchens, laundry, shower rooms, bathrooms, toilets, patient care tasks, etc. Review equipment and work practices and protocols for optimal ergonomics. Review logistics, work flow, storage placement of items, layout, organization of how duties are to be executed and their interplay with different departments, and how they may affect each other, etc.

Involve in-house physical therapists. They are the experts. Interview employees. They are the ultimate users. They are most qualified to determine what parts of the operation pose difficulties, and how best to resolve the difficulties with the most pragmatic solutions, usually. Add your own insights to the list. Refer to this list to conduct initial comprehensive ergonomic assessment. Utilize it later for periodic self-inspections for ergonomic hazards and potential improvements.

Job description review [basic]

Review job descriptions for areas of potential ergonomic risks that may warrant closer supervision, more training, and a continuing process for ergonomic improvement. Include the physical demands of the tasks necessary in the job description to ensure that applicants, and physicians who conduct fitness-for-duty examinations, can make informed judgments about their willingness and ability to do the job, respectively.

Near-miss reporting [optional]

A log of near misses should be kept to include incidents that could have occurred, or almost occurred. This is valuable information on existing hazards that can be acted on before an actual incident is produced as a result of the problem. This information should be pooled with information found through data review of the OSHA 200 Log and incident reports to enhance understanding of existing hazards, their priorities, and to realize their potential abatement solutions.

Videotaping [optional]

Videotaping of work in progress and work processes is very common as part of an ergonomic worksite analysis. It allows for repetitious review, freeze-frame, or frame-by-frame analysis of any movement or moment throughout a work process. Clearer analysis of actual and potential ergonomic risks can be made. It can also be a useful tool when counseling and/or training employees on the proper work methods.

Equipment Review [basic]

Equipment should be reviewed periodically, and before purchase, to ensure optimal ergonomics for the end users. Adjustments to existing equipment may be necessary, and sometimes replacement may be warranted. The purchase of new equipment is an opportunity to bring in good ergonomic design and prevent poor work designs. In conducting the ergonomic assessments, refer to ergonomic guides for optimal ergonomic design. *Always consult the end users before any major purchase or remodeling.*

Workstations, equipment, lift equipment, dollies, step stools, ladders, handheld tools, furniture, shelving, beds, transfer belts, lifts, sliding board, lift sheets, patient's clothing, employee uniforms, lifting space, floor area, shower chairs, etc. should all be assessed for potential risks they may pose to the users, and how their use may be improved. This involves assessing the entire job involved with each task. Review the size, shape, operation, method of use, place of use, transport, storage, etc. Review the placement or installation of shelving, equipment, or items, and the reach or body contortions necessary to utilize the equipment. Behavior aspects of work habits should be analyzed as well. For example, does one department overload certain equipment out of convenience, which may pose an ergonomic hazard for the other department that must deal with it later?

This assessment can be an ongoing activity at the ergonomic committee. It should begin with the more obvious hazards as seen from incident/injury data. As those are mitigated, the ergonomic committee can then systematically review other issues, or more persistent issues, over time. At some point, a comprehensive assessment of the facility will have been achieved, and the committee can streamline activities to a more maintenance-oriented program, with periodic ergonomic inspections, and focusing on new operations, or new issues brought up that had not been considered before or that warrant reconsideration.

Consult with in-house physical therapist, or other qualified experts [basic]

Consult with and utilize in-house physical therapists. They are the ergonomic experts. Fortunately, for long-term care, most facilities already have one, if not more. In the absence of this, or if this is not possible, enlist the help of another health care professional who is well versed in body mechanics, possibly someone in the in-service department. Unlike most other industries grappling with ergonomics, it is unlikely that long-term care facilities would need to pay an additional premium for such expertise.

Consult with employees [basic]

Also consult the employees. They are the ultimate users. Their feedback is very valuable in finding problems and the most convenient answers that will be *operationally pragmatic. Always consult with end users for feedback before purchasing any new equipment, or making any new changes to a work area or process.* Too often, consultants are brought in to cope with a problem created by a new change or "improvement." This usually results in make-do ergonomics abatement methods. Meanwhile, such situations could have easily been averted, or truly improved, if management, or the purchaser, had consulted the end users on the operational viability of the new piece of equipment or product purchased in the first place.

Always ensure that purchased equipment or product is selected not only based on pricing, vendor preference, warranty, service plans, etc., but also on user need, preference, and the potential risk of injuries to end users in the scheme of their *entire* job, not just the equipment, in and of itself. This relatively small investment of time, patience, and the cultivation of teamwork pays off greatly in the end. It reduces wasted time and money, and the likelihood of injury, while increasing morale and productivity.

Hazard Abatement [basic]

Interdisciplinary abatement methods or solutions will be developed to include consideration of better planning, organization, cooperation, communication, care planning, budgeting, personnel, assigned responsibilities, adequate expertise and authority, means to hold responsible persons accountable, teamwork, reinforcement, training, equipment, etc. These activities are interfaced with the safety committee (if the responsibility of the ergonomic program is meted out to a separate committee).

Follow-Up [basic]

Each hazard abatement intervention is followed up by an assigned individual within a set period of time to ensure abatement has occurred, and it is reviewed for effectiveness. When abatement does not achieve the desired or necessary results, the issue must be brought up for discussion. It should be determined why abatement was not successful. Efforts should continue in search of more viable alternatives and solutions.

Employee Training [basic]

Ergonomic training must be provided annually to all employees, if only to refresh them. Time spent for training and the scope of training should reflect the need for such time and scope based on accident/incident records and trends and observed behavior. Therefore, job-specific training outside the nursing department, which definitely requires patient transfer body mechanics training, should be provided as necessary.

Ergonomic training, both general and job specific, as needed, is provided to all new employees as part of their mandatory orientation. Job-specific training is considered to be given to *all* employees as necessary, and not limited only to nursing staff. Job-specific training will be provided to employees who are transferred to a new department or position, and when new equipment or processes are implemented. Formal ergonomic training will be provided at least annually. Instruction and feedback in the form of counseling or positive reinforcement from supervisors during work are ongoing.

Training on early identification, reporting, and conservative treatment of cumulative trauma disorders is provided to all workers, supervisors, and managers. Training material shall be presented in language and level of understanding appropriate for staff being trained. Training should be designed and implemented by qualified persons. Ergonomic training and educational programs for *all* general employees consist of the following:

- Explanation of the facility's Ergonomic Program
- Overview of the potential risk of back and other musculoskeletal injuries
- Causes and early symptoms of back and musculoskeletal injuries
- Means of prevention and proper lifting methods
- Treatment
- An explanation of the medical management procedures, and any alternative duty, and postinjury return-to-work programs
- An explanation of early recognition of the signs and symptoms associated with exposure, procedures to follow, and when to report signs and symptoms
- Staff physical fitness is encouraged.

Job-specific training [basic, if necessary based on record and observable practices]

Employees receive training from their own department on the measures employees can take in their particular setting to reduce ergonomic risks, on the correct use and adjustment of the workstation or area; on fixtures that will prevent or minimize exposure; and on proper work protocols to reduce ergonomic risks to themselves a co-workers. The depth and frequency of training is dependent on the need as seen from incident/injury record and observed work behaviors.

Nursing department job-specific training [basic]

The orientation/new employee program includes job site evaluation of transferring technique by a qualified person. Feedback is provided. Basic training in handling patients is given. Supervising nurses are trained to develop policies and procedures for their units and to train their nursing staff:

- Never transfer patients when off balance.
- Avoid heavy work with spine rotated.
- Lift loads close to the body.
- Avoid vertical "dead-lifts."
- Never risk overexertion with a patient who is resistant; use assistance.
- Use team lifts and mechanical devices where necessary (patients over 150 lb are always considered "heavy").
- Properly place and adjust equipment used to handle patients.
- Always bring patient toward you, never away.
- Do not lift fallen patient alone; get help, consider mechanical assistance.
- Employees (nurse's assistant, LPN, RN, etc.) are also provided:
— Proper training for assisting walking patients (stand on weaker side, close to patient, take load on hip rather than on back).
— Training in appropriate transfer and lift techniques in confined spaces, such as shower stalls and toiler areas.
— Training in when and how to use mechanical hoists (i.e., for patients over 150 lb who cannot support their own weight).
— Initial/periodic (i.e., monthly) programs on back care and transfer techniques.
— Training on the appropriate patient handling devices to use with particular patients and the proper use of this equipment.
- Training in emergency patient handling is given, such as for patients who have fallen, have spasms, are combative, or exhibit any unpredictable behavior.
- Training and feedback are provided on use of patient handling devices.
- Annual refresher training is provided and addresses specific needs. Provisions are made to train absent employees.

Dietary department job-specific training [optional]
Back injuries in this department result from transferring full steaming pots, pulling loaded carts, materials handling in storage areas, and slips and falls. Training should include:

- Proper storage of supplies and equipment (larger, bulkier items to be stored at waist level within easy reach, or lower, and lighter smaller items above the shoulder).
- Proper pot transfer and when to get help.
- Vigilance in keeping the floors clean and dry as much as possible.
- Proper body mechanics in stocking and retrieving materials in storage and freezers.
- Proper body mechanics in loading and unloading trays.

Laundry department job-specific training [optional]
- Proper storage of supplies and equipment (larger, bulkier items to be stored at waist level within easy reach, or lower, and lighter, smaller items above the shoulder).
- Vigilance in keeping floors clean and dry.
- Proper use of carts, especially in moving and unloading.
- Simple accommodations (step stool) to create better working levels for folding tasks.
- Proper work practice in transferring and moving large quantities of product (fluids).
- Proper body posture in loading and unloading washers and dryers.

Housekeeping department job-specific training [optional]
- Proper storage of supplies and equipment (larger, bulkier items to be stored at waist level within easy reach, or lower, and lighter, smaller items above the shoulder).
- Vigilance in keeping all floors clean and dry.
- Proper use of carts.
- Proper body mechanics, posture, and use of equipment in reaching difficult areas to clean (under beds, far side of tubs, etc.).

Maintenance department job-specific training [optional]
- Proper storage of supplies and equipment (larger, bulkier items to be stored at waist level within easy reach, or lower, and lighter, smaller items above the shoulder).
- Proper use of work tools including handgrips.
- Adjustments that may be made on workstations for better vision and comfort.
- Proper body mechanics and postures in completing tasks in awkward places and positions, including climbing ladders, crawl spaces, etc.

Office ergonomics [optional]
- Proper storage of supplies and equipment (larger, bulkier items to be stored at waist level within easy reach, or lower, and lighter, smaller items above the shoulder).
- Adjustments that may be made on workstations.
- Stretches and adequate breaks.
- Office safety (open drawers, looping cords, etc.).
- Proper body mechanics and posture in retrieving supplies, sitting at the desk, and fixing the office machines (copiers particularly).

Management employee training [basic]

Managers, supervisors, and other group leaders receive training comparable to that of all general employees in order that they can set a role model and identify proper and improper work procedures, and address both with counseling or positive reinforcement.

Employee Health [optional]

Encourage staff physical fitness (weight loss, exercise, stretching, healthy lifestyles).

Preventative Maintenance and Good Housekeeping [basic]

An effective program for facility and equipment maintenance is implemented that minimizes ergonomic hazards. The program includes a preventative maintenance program for patient handling devices; maintenance whenever employees report problems; existance of sufficient spares for out-of-service equipment; implementation of housekeeping programs that minimize slippery work and surfaces and slips and falls; etc.

Nursing Department Ergonomics Management [basic]

Changes are implemented to avoid the most stressful patient transfers. After each patient handling injury, a determination is made as to whether a task can be modified to reduce future risk. Ergonomic hazards are prevented primarily by proper selection of equipment, effective use of assistive devices, and implementation of proper work practices. The equipment must be available in sufficient quantities, convenient for use, and properly maintained.

- Patient chairs that allow the nurse to place feet under chair, chair back is low enough to permit lifting access.

- Mobile chairs, wheelchairs, and commodes have functional brakes, mobile arms, and footrests that do not obstruct movement.
- Transfer surfaces are approximately at the same level; for example, toilet seats are raised so that the heights of the wheelchair and toilet seat are the same. Bath stretchers are height adjustable to allow horizontal transfer.
- Hoists are selected based on nurse and resident evaluation of ease of use, comfort, and safety. Hoists are routinely maintained, and conveniently located and available, with suitable lifting attachments and/or slings, such as:
— Fixed hoists often used in bathrooms due to limited space.
— Hoists on a track, which can also be operated by patients.
— Mobile hoists.
- An adequate number of hoists are available for transfer of residents who are dependent or can only provide minimal assistance in the transfer.
— At least one portable resident hoist with a maximum capacity is available for extremely heavy patients. A backup hoist is available when the primary lift is out of service.
— One hoist, with an adequate number of slings, is available in each wing and easily accessible.
— At least one of the portable hoists in the nursing home should have the functional capacity to lift a resident from the floor.
— A digital scale is attached to one mechanical hoist to allow the weighing of dependent residents during routine transfers.
- Guidelines have been developed to instruct nurses in which type of hoists or sling would be appropriate for particular patients.
- Nursing supervisors are trained so that they understand the importance of using patient handling devices and the techniques for using them.
- "Sliding boards" are used to ease transfers from wheelchair to bed and bed to wheelchair when the resident has the cognitive and upper body strength to assist.
- Transfer belts are used where appropriate. (Choose wider cushioned transfer belts with multiple handle straps and quick release latches.)
- Patient transfers are minimized. Shower chairs are utilized, which are also compatible with toileting facilities. Weighing stations allow residents to remain in their chairs while being weighed on a wheelchair ramp.
- Good ergonomic bed design is evident through convenient location of controls, adjustable bed heights to wheelchair, and sufficient foot clearance.

Work Practices and Administrative Controls

- Supervisors are familiar with patient handling guidelines and enforce facility rules. Care plans are specific in assessing patient handling requirements, and are communicated to affected employees prior to patient handling. A practical system for communicating changing assessment results is developed, and special attention is given for residents who have special handling needs, such as residents who have weakness on one side from a stroke.
- Patient clothing is selected that aids transfer feasible (nonslippery; clothing handles for the unconscious residents; adaptive clothing for ease of changing; absorbent pads for incontinent patients).
- Proper patient transfer techniques are utilized (patient is properly positioned prior to transfer; patient handling slings or transfer belts are utilized; patient clothing is never adjusted while transferring).

- Slips and trips will be minimized by leveling uneven floor surfaces, using large wheels on transport equipment, creating nonslip surfaces in toilet/shower areas, and enforcing a policy of immediately cleaning up fluid spilled on a floor.
- An existing policy ensures that patient handling help is provided to a nurse any time it is requested.
- Restricted "alternative-duty employees" are not involved in patient transfers.

Incentive Programs [Optional]

Ergonomic and safety performance evaluations are correlated to benefits, incentives, or are otherwise recognized through a formal recognition or safety incentive program. Recognition or incentive programs must be based on observable *desired actual performance*. Recognition or benefits are not awarded based only on the lack of incidents or injuries, which may very well not be associated with actual good ergonomics and safe behavior, but merely on the random nature of risk.

Medical Management Program [basic]

Program administrator
The program is supervised by a person trained in the prevention of cumulative trauma disorders, and is a consultant to the Safety Committee.

Preemployment and annual physicals
Preemployment and annual physicals and medical screening are conducted in a manner that does not violate the Americans with Disabilities Act, but that assesses the applicant for fitness for duty. This is achieved by the composition and use of well-written job descriptions that delineate all the physical (mental and emotional) demands of the job that are "essential job duties" and "nonessential job duties." Inquiries shall not be made about specific disabilities, or past workplace injuries, which is illegal.

Applicants or employees will be evaluated based on their *ability to meet the requirements of the job*. Questions and examinations shall be based on the ability to do the job, and what accommodations, if any, are necessary to do them. If the applicant or employee is not able to meet *even one* of the "essential job duties," then the person is not fit for duty, or qualified for the job. However, accommodations must be made, even if the applicant or employee is not able to do *any* of the "nonessential job duties."

Accurate injury and illness recording
This program assesses the accuracy of the injury and illness data recorded. It is the driver of the reporting system for this type of information. It is the responsibility of this program to ensure that incidents are reported immediately and follow-up is conducted as quickly as possible.

Early recognition and reporting
Early recognition, evaluation, treatment and recovery of musculoskeletal disorders related to overexposure to ergonomic risk factors at work is facilitated by this program via employee training, symptom reporting, implementation of interventions, monitoring of symptoms, referrals to appropriate medical treatment, particularly early conservative treatment. All employees shall be provided with information regarding the signs and symptoms of musculoskeletal disorders and encouraged to promptly report early signs and symptoms to the facility_____Department. All injuries and illnesses reported that are associated with ergonomic risk factors shall be documented including the following:

- Date the symptoms were reported
- Location, type, and severity of symptoms
- Description of work activities that were being performed when symptoms began

- Preexisting medical conditions associated with the employee
- Non-work-related activities and injuries
- Prior treatment related to current symptoms

This information shall be continuously monitored and analyzed for trends relating to ergonomic risk factors in the workplace. All personal information documented shall be kept confidential. No employee shall be subjected to discrimination based on such reporting.

Such reporting shall allow for timely evaluation and intervention. Not all employees with reported symptoms warrant immediate evaluation by a health care provider. If risk factors are determined and symptoms persist with the employee, he or she shall be referred for an evaluation by a qualified health care provider.

Intervention, monitoring, and return to work

Depending on the direction of the health care provider, employees shall be given sufficient time for the involved muscle, tendon, or nerve to heal. Interventions shall be implemented that are likely to be beneficial for symptoms reported. Exposure is reduced through job rotation, placement into the Alternative Duty Program, work restrictions, splints and supports, etc., if available. Gradual resumption of duties over time will be allowed if possible. The facility may develop a list of jobs with the lowest ergonomic risk derived from job ergonomic analysis. This list of light-duty jobs shall be periodically reviewed and updated. All employees assigned to restricted-duty jobs shall be placed in appropriate jobs consistent with their capabilities and medical limitations. There should be periodic assessments to increase employee job function with increased fitness throughout the rehabilitation process.

These alternative or restricted duties are only temporary assignments throughout the rehabilitative process. This period should be established and limited to, preferably, no more than 1 to 4 months, depending on the particular situation (FMLA policy, termination and disability policy, union contract, labor pool, management style, quality of health care providers and level of cooperation, etc.). If the employee is not capable of returning to full duty as per the job description by the set time limit, then they are no longer fit for duty, or qualified for that job. Employees must have medical clearance to return to full duty.

Periodic evaluation of program

This program is evaluated periodically.

Back Belt Policy [optional, but really basic]

Braces, splints, back belts, and other similar devices are not personal protective equipment. There is no conclusive evidence regarding the effectiveness of such support devices. Many studies have suffered from poor data collection and interpretation. There are three positions to choose from. At the time being, the authors recommends the first. The other two position are presented to provide guidance for those clients who choose the alternate policy. All policies should be reviewed periodically to assess the appropriateness of the position in light of any new data. Whatever the facility position, it should select one, express this in the form of a written policy here. This policy should be communicated to employees and continually implemented.

Policy 1

There is no conclusive evidence on the effectiveness of back belts in preventing back injuries. Although some studies have had positive results, their financial sponsorship and scientific validity have been questioned. Few studies have been designed to rule out additional contributing factors, such as concomitant safety efforts, or the "Hawthorne effect" (safety effort placebo). Meanwhile, there are questions about the potential problem of muscle atrophy and the "Herculean effect." Employees should be medically cleared to wear a back belt, as it does induce stress on the pulmonary system and increases blood pressure. Therefore, in view of this and on the strength of other more reliable interventions, we are discouraged from recommending the economic investment and unknown risks of implementing a back belt program, and would recommend that medical clearance be required for any back belt wearers.

There is no back belt use in this facility without appropriate medical documentation, with a specific diagnosis, prescribing the specific support device, and length of time that it is to be worn. Back belt training will be provided to employees to explain the facility's position on the device.

Policy 2

There is no formal back belt program. Employees will not be furnished with the devices. However, employees who wish to wear them are allowed to wear them on the premise while working. Employees must be medically cleared to wear one.

Policy 3

The facility chooses to implement a back belt program. Back belts are mandated and provided to all employees who do lifting. Employees must be medically cleared to wear one. Ergonomic training on proper body mechanics is provided along with training on the proper use of a back belt, its intent, and potential drawbacks.

Ergonomic Policy Review [basic]

This written program is reviewed and updated at least annually and whenever necessary to reflect new or modified tasks, programs, and procedures that affect employee exposure to ergonomic hazards.

Reviewed by:

_____	_____	___/___/___
Name	Title	Date
_____	_____	___/___/___
Name	Title	Date
_____	_____	___/___/___
Name	Title	Date
_____	_____	___/___/___
Name	Title	Date
_____	_____	___/___/___
Name	Title	Date

III. TUBERCULOSIS

A. GOAL

In an effort to curb the potential spread of TB among health care workers, health care employers are required to review and update their TB Infection Control Program.

B. SUMMARY

OSHA published and distributed an enforcement guideline (February 9, 1996) to its inspectors to follow, in the event they should conduct a health care inspection. There are, of course, the Centers for Disease Control (CDC) revised guidelines as of June 7, 1997.

PPD testing should be performed before employment, and annually thereafter, for a baseline test. The test should be the Mantoux skin test, not the old PPD four-pronged prick test. Conversions on baseline tests (not upon initial employment) are considered to be recordable on the OSHA 200 Log as an illness. Positive PPD resulting from a booster test are not recordable on the OSHA 200 Log.

It may be advisable to perform a booster PPD test on incoming employees who cannot verify a recent PPD test, or TB-positive status, particularly, depending on the community risk profile. Otherwise, antigens may be so low in a truly PPD-positive person, that they may result in a false negative on the initial PPD test. But the next year, when their antigens are back up to detectable levels because of the PPD booster, their annual baseline test is then documented as a conversion. When this occurs, the case must be entered onto the OSHA 200 Log, misconstruing an initial false negative for a conversion due to a potential alleged exposure at the facility. This scenario serves as a flag and the inspector must conduct a more *focused inspection* on the facility's TB management. On a highly unlikely, but entirely possible note, a latent TB case may develop into a TB workers' compensation claim, which never originated in your facility but which you will have to controvert and, one hopes, win. In addition, PPD conversions require appropriate epidemiological reviews and in-house infection must be ruled out and well documented.

There are specific employee training requirements on TB, just as are delineated for bloodborne pathogens. There are also work removal, medical screening (both employees and residents), and transportation of TB/suspect TB patient requirements (see *Synthesis Summary* in Section 4 III.F). It would be practical and prudent to have well-thought-out and well-planned contingency plans for a variety of potential TB situations including resident(s) or employee(s) with suspect TB, or PPD conversion; health care providers disagreement on suspect or nonsuspect status of a resident; the receiving facility not prepared or willing to admit the facility's suspect TB patient; etc.

C. UPDATE

Health care worker lobbies are pressuring OSHA for a TB standard. Most of the requirements and rationale for requirements in the enforcement policy are based on CDC information. As CDC information and recommendations change, so will the expectations of OSHA.

On the surface, a policy stating that the facility will not handle any suspect or active TB patient will suffice. The policy must contain procedures on immediately discharging the patient to a facility with negative pressure capabilities. The policy must also specify that the patient will not be readmitted unless three negative sputum smears over three consecutive days have been obtained as per CDC recommendations.

Note: Be wary of employee interviews that may result in opinions that such a policy was not indeed followed. There may have been instances when employees believed the resident to be suspect or that they were at risk, but the facility did not respond, as it should have. Such concerns would most likely be aired during OSHA inspection interviews with employees. Ensure that the facility actually follows its own policy strictly, and that such procedures are well documented. For further guidance on what to do in such cases, or to review a facility's risks of such inadvertent incidences, please refer to the Practical Guide in Section III.E as discussed below.

D. CONTENTS

The first document, a practical guide for OSHA compliance and TB management in long-term care facilities (in Section III.E), is a hands-on discussion to help facilities *determine their real risks and needs.* It provides them with the basic tools to begin some preliminary recommended actions *for those who deem it appropriate and necessary* for their operations.

The second document, a synthesis summary for long-term care facilities (in Section III.F), is an interfaced guide on OSHA inspection protocols on TB and the CDC guidelines (edited to include only information that relates to nursing homes; items concerning hospitals, homeless shelters, AIDS clinics, prisons, and other medical clinics have been deleted). This document is applicable to *all* facilities regardless of their individual TB management strategy.

These two documents are designed for facilities who choose to take proactive measures to ready and protect themselves from the shortfalls of the guidelines discussed above, while affording themselves the luxury of control and conservatism should any question of suspicious TB symptoms arise.

The section closes with important TB respirator notes (in Section III.G) and a sample written respiratory protection program for long-term care as part of the TB infection control program (in Section III.H).

E.   PRACTICAL GUIDE FOR OSHA COMPLIANCE AND TB MANAGEMENT IN LONG-TERM CARE FACILITIES

This guide provides sound, reasonable, and economically feasible methods to manage TB incidents, in order to minimize disruptions to operations, employees, and residents; ensure health and safety; and to expedite care and service.

Introduction

TB policies in most long-term care facilities typically state that the facility will not handle a "suspect TB" case. Any suspect cases will be removed from the facility, and not returned until cleared according to CDC guidelines. This may be all well and good on paper. It will even pass muster with OSHA and the DOH. However, it does not prepare facilities for a "TB incident," "TB problem," or "TB question," which are more probable than an all-out "suspect TB" diagnosis.

Generally, these scenarios will not become a problem if a facility has extremely good labor relations and medical direction. Nonetheless, facilities have even more to gain if steps are taken to enhance their ability to work with a questionable or developing TB situation. Keep in mind that the agency mandates, of CDC, DOH, and OSHA, are to protect employees, the public, and residents. They neither focus on nor consider the provisions necessary to cover the practical needs of employers.

The Problem

There are gray areas where there is a question of the status of a case being "suspect TB," or not. Health care providers exercise discretion and discrimination in many cases. Doctors vary to the extent of how conservatively they will make a definite diagnosis, for a number of reasons. It seems that those physicians who can afford to be conservative (very cautious), can be overtly so. Residents' physicians, or medical personnel in a long-term care facility tend to be more conservative about making such judgments. They have little to lose, and everything to gain, including safe assurance. Sometimes, for a resident's private physician, financial compensation is improved if the resident is moved to a hospital.

Hospital personnel, on the other hand, tend to exercise more discerning judgment since caring for a patient in isolation incurs costs, risks, and problems. The hospital incurs risks of exposure to employees and other patients, especially those with HIV. It will require the use of isolation rooms and procedures; however, it produces little revenue in terms of "billable" medical procedures or laboratory tests performed, given the long stay of the patient. Some hospitals actually lack the facilities; others may be reluctant to have to utilize them.

We have coined our own terminology, *suspiciously suspect*, as sometimes there are conflicting medical and lay judgments about the suspect nature of a case. Existing blanket TB policies on removing suspect TB residents and not taking them back until they have been cleared do not provide facilities with many options, or much flexibility, in dealing with TB situations as they arise.

The Situation

Conflicting medical opinions on whether to remove a resident to a hospital or to discharge a resident early from hospital isolation (before the three negative sputum smears in three consecutive days) creates a conflict between the hospital and the nursing home, as well as confusion and panic at the nursing home. The disruptions to the operations, employees, and residents are usually brought on by the hysteria that arises when it is generally perceived that the facility is harboring an individual whose TB status is not yet known, or has been improperly disclosed. They also occur when a facility consents to readmitting a resident prematurely, despite its own policy. This results in employee anxiousness that, at least, affects productivity and quality of care and, at worst, leaves the facility vulnerable to confrontations with the union, OSHA, DOH, and/or the local news media.

In many cases, the risks a facility faces with a TB question are not so much a health problem, although it leaves much to be desired to lack a solid contingent plan, but one of containing and managing employee hysteria and public opinion regarding the degree of control a facility maintains, and the appropriateness of its actions. This often can equal, if not surpass, the urgency of the issue of a health exposure. *Strong medical justification to mitigate inquiries and assuage hysteria is a reasonable argument.* The problem arises when the medical evidence is weak, conflicting, or unconvincing. In such cases, expressed concern is founded.

Some Real-Life Scenarios

There have been situations where the hospital is unable, or unwilling, to take the resident, or there is a delay. This is also the time where hospital medical personnel may deem the resident not to be a suspect TB case. However, a facility's own medical director or the resident's physician may decide that the case is indeed suspect and that the resident should be in isolation. On the other end, facilities have been strongly pressured to take back the resident, even before three negative sputum smears from three consecutive days have resulted. This becomes a game of hot potato, on the basis that the potato is not really hot.

Meanwhile, employees will be upset and nervous. They will sense that there was no contingent plan for such incidents. They may think that administration or management is dealing with it on a crisis basis. In most cases, this is, in fact, true. Unions are notified, OSHA is brought in. Pregnant employees demand to leave, or are dismissed by confused and alarmed supervisors. Employees fear that they will be exposed to TB because the facility did not have clear guidelines, and is tolerating and allowing *less than optimal risk exposure* because it does not care, and/or that external forces beyond its control are adversely affecting them.

Nonetheless, even if you do not foresee that you will experience such problems, the following practical recommendations would provide you with the assurance and ability to care for residents, with suspect TB, at the beginning or end of their diagnosis, and to provide an additional standard of safety for the staff.

The Cure

Fortunately, good management and good medical direction will prevent such issues from arising. There are some very simple but fundamental steps that can be taken to integrate with your existing policy, to avoid such a situation. This guide will assist you in appearing and remaining diligent and in control, with minimal effort. It will even provide you with more security, flexibility, and options in keeping a resident who is "suspiciously suspect" or actually "suspect."

OSHA Requirements

To meet basic OSHA *expectations to recognize and abate potential TB hazards* (as there are no real specific requirements as would be mandated by a TB Standard), be prepared to document and implement the following:

- Maintain a TB policy stating that suspect TB residents will be removed from the facility, and that the facility will not care for any such residents, until such time as they are cleared according to CDC guidelines.

- Implement medical screening—Annual PPD (Mantoux) test for employees.
- Record any conversions from a baseline negative on the OSHA 200 Log on the Illness Section of the form (not for new hires).
- Monitor and investigate each PPD conversion, and signs and symptoms of TB among employees and residents.
- Provide for work removal of infectious employee.
- Implement annual employee TB training.
- Fit TB patients to be transported for evaluation with an appropriate mask. If feasible, the resident shall wear a mask during transport. Arrange for transporters to wear an appropriate respirator when transporting resident to hospital.
- Require negative-pressure respiratory isolation rooms for suspect TB residents.
- Assure that employees working in negative-pressure isolation rooms with suspect TB residents wear appropriate respirators as guided by a respiratory protection program, which includes initial fit testing, routine fit checking, equipment maintenance, and annual employee training.

Practical Guidance

The following steps are recommended to provide a long-term care facility with forethought and planning to safely and comfortably deal with any type of TB question, issue, or incident that may arise. It will help to expedite resident removal and readmission to and from a hospital.

Preparation for a temporary isolation room, and the ability to run one simply and well, allows the facility to "play it safe" much easier. It provides a facility with more options, flexibility, and abilities to care for residents with a weak to strong probability of being suspect TB, without involving external entities or arousing doubts among staff and family.

Administer booster PPDs to appropriate new hires. For the facility's sake, it is worth the extra effort to ensure that an initial baseline negative is not a false negative. Remember, conversions from a negative baseline are recorded on the OSHA 200 Log, not to mention any potential workers' compensation claim liability should conversion, or ensuing illness, occur coincidentally with a TB issue, question, or situation at your facility. Consult your infection control nurse or medical director for more details.

Formalize your in-house TB policy with neighboring hospitals. Ensure that your neighboring hospitals are aware of your policy, and agree to cooperate with you in the event that a suspect TB resident requires hospital admission.

Establish how consensus on medical diagnosis and prescription for negative-air-pressure isolation will be achieved between the facilities. Ensure that these decisions are made with input from all essential medical personnel involved from both facilities. Protocol must be established beforehand, in the event that there is conflicting medical diagnosis and indications. (Thus, the recommendation for a simple but effective temporary negative-air-pressure isolation room.)

Be cognizant of the general airflow patterns of your facility. It makes sense to be aware of the basic airflow patterns in your facility, as part of the management of TB and any other nosocomial airborne infectious diseases. You can also determine the optimal room to designate as the temporary negative-pressure isolation room. You will know if you can easily isolate residents in their own room, or if you need to consider moving them to one better suited, that can be made available. Awareness of airflow patterns may point your infection control program to understanding and evaluating this facet of the spread of other infectious diseases in the facility.

Practical Note: A small box of ten smoke tubes costs about $50. This should last you for a few years, depending on use. These are used to detect air movements. The maintenance department, in conjunction with infection control, can determine general airflow patterns. Keep a record of their findings, in the event of a suspiciously suspect or suspect TB case, or a case of any other relevant airborne infectious disease that may find this information helpful. Refer your maintenance or environmental services personnel to Field Guide 1: Determining Airflow Pattern for smoke tube product and ordering information, as well as implementation guidance and details.

Make provisions to make the optimal room available in the event you need it. If the best room happens to be a private room, facilities have created a policy where incoming private residents in that room agree to, and understand, that in the rare event of an isolation situation they will transfer to a temporary room or bed until isolation is no longer necessary.

Make provisions to make the optimal room a temporary, emergency negative-pressure isolation room. This is really not as difficult as it sounds. All you need to do is seal off a room and provide relatively basic negative pressure in the form of a small window fan or, in some facilities, a bathroom fan that vents directly to the outside. Refer to Field Guide 2: Creating a Temporary, Emergency Negative Pressure Isolation Room for detailed directions and suggestions. This section also includes simple worksheets on calculating air change rates of rooms. Six air changes per hour are required in a respiratory isolation room.

Testing and activating the temporary, emergency negative-pressure isolation room. Once the negative-pressure isolation room is set up, use the smoke tubes to verify and confirm that you can, and have, achieved negative pressure in any room you have sealed off, as necessary. Refer to Field Guide 1: Determining Airflow Pattern negative-air-pressure testing protocols. When the room is being used for respiratory isolation, airflow should be checked and documented daily.

Respiratory Protection Program. Any employees working in the activated temporary negative-pressure isolation room should have an appropriate respirator, and be covered by a Respiratory Protection Program. OSHA requires a Respiratory Protection Program, even if only one employee wears a respirator. The program must be written, and provide for medical clearance for employees' ability to wear a respirator; respirator fit testing (which the author's office is qualified and able to perform for you, if your manufacturer or supplier is not willing or able, or if you simply prefer us to); proper respirator selection; proper respirator maintenance, storage, and use; and annual employee training. Please refer to the *sample written respiratory protection program for long-term care* (in Section III.H).

Practical Note on Staff Selection: Carefully select key staff from each shift and each department as necessary to cover all of the resident's needs for a prolonged period of time. Obviously, long-time staff should be favored over new hires to reduce potential redundancies with frequent turnovers. (You will need to be job specific if you have a union. If not, you have more flexibility as selected staff with respirators can perform a variety of duties while in the isolation room.) Further, explain that individuals who are HIV positive, or who are immunosuppressed, are at greater risk of becoming ill with TB, and possibly dying in the extremely rare event that they contract multidrug-resistant TB. Allow selected staff to decide if they will take on such assignments (as opposed to mandating them), after they have considered the risks and their own personal medical status.

Practical Note on Spin-Off Uses of Respiratory Protection Program: Here is an opportunity to legitimize any other respirator use in the facility, including respirators used by maintenance in spray painting, or other such activities. Such users should be included in the Respiratory Protection Program. Other facilities have declined using respirators for such tasks, but would have liked to offer the added protection, if not for the inconveniences of having to set up a respiratory protection program. Now the program is in existence, the impetus is there to provide that added degree of protection to the maintenance staff.

FIELD GUIDE 1
Determining Airflow Patterns—Technical Guide
Materials

- Blueprint, map, or sketch of facility layout
- *Smoke tubes* for testing air ventilation. These can be ordered through any supplier of occupational health and safety products. (An example is Lab Safety Supply at 800-356-0783 whose airflow kit with six smoke tubes with over 100 tests each, aspirator bulb, and carrying case costs about $66, order 7#-23060, smoke tube reorder tube 7#-9236-2 for 10 tubes at $33.)

Directions to Identify General Airflow Patterns

1. Break off both ends of smoke tube.
2. Place one end in bulb.
3. Systematically walk through the facility and squeeze bulb at specific locations noted on the facility map. Puff smoke at breathing level, floor level, and high above your head. Record readings using appropriate demarcations for each location where readings were given (examples, H for high over the head, M for medium height, L for floor level). Mark direction of airflow, and velocity as strong, weak, or none.
4. Note season, time of day, temperature, humidity, weather conditions.
5. Repeat for each season.
6. Determine general airflow throughout the facility.
7. Identify the last resident room in the airflow stream (on each floor).
8. Identify the resident rooms in the end of the airflow streams.

These rooms should be considered for temporary negative-air-pressure respiratory isolation rooms, if possible. Resident rooms at the beginning of the airflow patterns should be avoided.

Directions to Evaluate Negative-Pressure Room

With the door closed, and all air control systems on, puff the smoke tube at the bottom of the isolation room door, a few inches away from the door. The smoke should be sucked under the door. If not, there is inadequate negative pressure.

Also, in the isolation room, with the door closed, and the air handlers on, test the air pattern in the room to ensure that the air does move toward and out the exhaust.

FIELD GUIDE 2
Creating a Temporary Emergency Negative-Pressure Respiratory Isolation Room

Directions:

When a suitable room has been established as per Field Guide 1:

1. Make certain that all recirculating vents are disabled, and sealed off (duct tape).
2. Seal off all other potential air leaks (electrical receptacles, exhaust vents, etc.).
3. Determine that air change rate is adequate as determined through calculations below.
4. Keep bathroom exhaust fan on at all times if available, or keep window fan on at all times as necessary.
5. Maintain the airflow and air rate at all times while isolation room is in use.
6. Check for airflow daily, and document daily when isolation room is in use.
7. Isolation room door should remain closed at all times and be marked with a sign, "Authorized Personnel Only" and/or "Respiratory Isolation Please Keep Closed at All Times."
8. If placing a window fan into the window for added ventilation, provide for a good window seal, if possible, as necessary to result in the level of negative pressure needed.
9. Sequester the area 8 feet beyond the outside of the window to minimize inadvertent exposure to passersby.
10. Calculate for number of air changes per hour in respiratory isolation room (the information in bold is the calculated value). The information underlined is the measured value.

Room Number:_____

1. **Room Size in cubic feet** = length × width × height
2. Area of bathroom exhaust fan in square inches = length × width/144 (conversion factor to change to square feet) = **area of bathroom in square feet**
3A. Airflow measured with a velometer in feet per minute =

or

3B. Rated airflow of fan (information is supplied by the manufacturer) in cubic feet per minute =

4A. **Airflow in cubic feet per minute** = measured airflow (from 3A) × **area of bathroom exhaust fan** in square feet (if using velometer above in 3A to measure airflow)

or

4B. **Air Flow in cubic feet per minute** = rated airflow × **area of bathroom exhaust fan in** square feet (if using rated airflow from manufacturer above in 3B)

5. **Airflow in cubic feet per hour** = airflow in cubic feet per minute × 60 minutes
6. **Air changes per hour** = airflow in cubic feet per hour divided by room size

If the air changes per hour is fewer than 6, the room cannot be used for respiratory isolation. Make addition air handling changes as necessary, to achieve a minimum of 6 air changes per hour.

F. Synthesis Summary for Long-Term Care Facilities of OSHA Directive on Enforcement Procedures for TB Exposure and CDC Guidelines for Preventing the Transmission of TB in Health Care Facilities

Introduction

The following is a summary of the final version of the (1996) OSHA TB Enforcement Directive as it relates to *long-term care facilities*. Relevant parts of the latest CDC Guidelines for Preventing TB Transmission in Health-Care Facilities (1997) as referenced in the OSHA TB Enforcement Directive is also included herein.

Copies of these documents in their entirety are available from the respective agencies. This synthesis is an attempt to distill the essential elements of both. All components of the OSHA TB Enforcement Directive that have been excluded are not relevant to long-term care or, at least, the average long-term care facility. However, many components of the CDC TB Guidelines for Health-Care Facilities that may not be covered or included in greater detail here, can serve as useful references (e.g., drug therapy recommendations, and other more clinical aspects of handling and treating suspect or confirmed TB cases on an in-patient basis).

Some terms have been altered that do not change the integrity of the meaning but that are more familiar to the long-term care audience. Also, some more specific references have been interjected to make the guidelines more relevant.

OSHA Inspection Scheduling and Scope

1. *Employee complaints* can lead to an OSHA TB inspection.
2. Related fatality or catastrophe can lead to an OSHA TB inspection (TB outbreak, a multidrug-resistant strain case, or a perception thereof).
3. A TB inspection may be part of all industrial hygiene inspections conducted in workplaces where the CDC has identified workers as having a greater incidence of TB infection than the general population. Long-term care facilities have been subject to reports issued by the CDC providing recommendations for TB control and thus are included in this list.

Inspections shall include a review of the employer's plans for employee TB protection, if any. Such plans may include the infection control program, respiratory protection program, and skin testing. Employee interviews and site observations are an integral part of the evaluation process.

OSHA Inspection Procedure

Upon entry, the inspector shall request the presence of the infection control director and employee health professional responsible for occupational health hazard control. Other individuals who will be responsible for providing records pertinent to the inspection may include in-service director, facility engineer, director of nursing, etc.

The inspection shall establish whether or not the facility has had a *suspect* or confirmed TB case *within the previous six (6) months* from the opening conference to determine coverage under the Occupational Safety and Health Act. This determination may be based upon interviews and a review of the infection control data.

If the facility has had a suspect or confirmed TB case within the previous 6 months, the inspector shall proceed with the TB portion of the inspection:

1. Inspector shall verify implementation of the employer's plans for TB protection through employee interviews and direct observation where feasible.

2. Professional judgment shall be used to identify which areas of a facility must be inspected during the walkthrough (for example, the affected areas or area in question, isolation room).
3. Compliance will be determined after review of the facility plans for worker TB protection, employee interviews, combined with an inspection of appropriate areas of the facility.

Inspector's Internal Protocol

1. Inspectors are supposed to be familiar with the CDC guidelines, terminology, and be adequately trained through coursework or field experience in health care settings. They are encouraged to consult with their regional TB coordinators prior to beginning such an inspection.
2. Inspectors are not to enter occupied respiratory isolation rooms to evaluate compliance unless, in their determination, entry is required to document a violation. Prior to entry, the inspector will discuss the need for entry with his or her area director.
3. Inspectors may use photographs or videotaping where practical for case documentation. Under no circumstances shall photographing or videotaping of patients be done. Inspectors must take all necessary precautions to assure and protect patient confidentiality. (Thus, explain your in-house resident confidentiality policy and procedures with the inspector to coordinate necessary arrangements for documentation while protecting resident confidentiality, as the need arises.)
4. Inspectors are to assume that the isolation room is not under negative pressure if an isolation room is occupied by a patient with suspect or confirmed TB, or the room has not been adequately purged, and shall wear a negative-pressure HEPA respirator upon air testing or entering the room if determined necessary.

Citation Policy

Whenever an employee may have an occupational exposure to TB, the following Standards must be complied with:

Section 5(a)(1)	General Duty Clause
29 CFR 1910.134	Respiratory Protection
29 CFR 1910.20	Access to Employee Exposure and Medical Records
29 CFR 1904	Recording and Reporting Occupational Injuries and Illnesses

Violations of these OSHA requirements will normally be classified as serious.

General Duty Clause

General Duty Clause Section 5(a)(2) states, "Each employer shall furnish to each of his employees employment and a place of employment which are free from *recognized hazards* that are causing or are *likely to cause* death or serious physical harm to his employees."

General Duty Clause citations shall be issued only when there is no standard that applies to the particular hazard. *The hazard, not the absence of a particular means of abatement, is the basis for a General Duty Clause citation.* All applicable abatement methods identified as correcting the same hazard shall be issued under a single General Duty Clause citation.

Note: The General Duty Clause is the catch-all citation for hazards that are not specifically covered under a particular standard. However, in order to be cited, the hazard has to be a *recognizable hazard, likely to cause* death or serious physical harm. There also have to be abatement methods *available* that the employer has not employed.

For purposes of citing Section 5(a)(2) for TB exposure noncompliance, *recognition of the hazard,* is shown by the CDC TB Guidelines for Health-Care Facilities for the types of exposures detailed in the following because the CDC is an acknowledged body of experts familiar with the hazard.

Citations for long-term care facilities shall be issued when the employees are not provided appropriate protection *and* who have exposure as defined below:

1. Exposure to the exhaled air of an individual with *suspected* or confirmed pulmonary TB disease. *Note:* A *suspected* case is one in which the facility has identified an individual as having symptoms consistent with TB. The CDC has identified the symptoms to be productive cough, coughing up blood, weight loss, loss of appetite, lethargy/weakness, night sweats or fever.
2. Employee exposure without appropriate protection in a high-hazard procedure performed on an individual with suspected or confirmed infectious TB disease which has the potential to generate infectious airborne droplet nuclei (e.g., *aerosolized medication treatment,* sputum induction, endotracheal intubation, and suctioning procedures).

Feasible and Useful Abatement Methods

The following are examples of feasible and useful abatement methods, which must be implemented to abate the hazard. Deficiencies found in any category can result in the continued existence of a serious hazard and may therefore allow citation under the General Duty Clause 5(a)(1):

1. Early identification of patient/client

The employer shall implement a protocol for the early identification of individuals with active TB. There is a reference here to pages 19 through 30 of the CDC Guideline which covers:

- Development of the TB infection-control plan
- Periodic reassessment
- Identifying, evaluating, and initiating treatment for suspect TB patients
- Management of institutionalized patients with confirmed or suspect TB (which includes initiation of isolation for TB, TB isolation practices, the TB isolation room, discontinuation of TB isolation, and discharge planning)

Note: Although most long-term care facilities claim that they will not hold any individual who is even suspect TB, the facilities must have clear policies on when an individual is a suspect TB case, and how firmly they will stick to their immediate discharge policy, and what the interim procedures will be to care for the resident as he or she awaits transportation, or further administrative decisions.

It is inadequate to have nothing except an understanding that all suspect cases will be discharged to a hospital. Facility policies must be clear enough to refute employee claims of suspect cases that were not properly identified or handled. It must also have adequate procedures to care for the resident while awaiting transportation to the hospital, or in the event that the resident cannot be moved for one reason or another.

2. Medical surveillance

- *Initial Exams.* Initial TB skin tests to all current potentially exposed employees and to all new employees prior to exposure. A two-step baseline shall be used for new employees who have an initially negative PPD test result and who have not had a documented negative TB skin test result during the preceding 12 months.

Note: The two-step baseline rules out false-negative results of new employees. It is also beneficial to the facility since failure to do so, not only is noncompliant with the CDC recommendations, but also may inadvertently place the liability of a TB conversion on the facility a year later when the false negative is boosted to reveal itself as a real positive. Such a case is recordable on the OSHA 200 Log. It could have other ramifications on a facility's workers' compensation claims management should the case develop into disease. Thus, it is doubly worthwhile for the employer to make certain that the initial PPDs are accurate to avoid liability for previously existing TB infections that did not occur at the facility.

- *PPD Interpretation.* The reading and interpretation of the TB skin test shall be performed by a qualified individual who has a clear understanding and knowledge of the clinical interpretive aspects of PPD reactions.
- *Periodic Evaluations* shall be conducted annually for low-risk personnel (most long-term care facilities have assessed their employees as low risk). Workers with a documented positive TB skin test who have received treatment for disease or preventive therapy for infection are exempt from the TB skin test but must be informed periodically about the symptoms of TB and the need for immediate evaluation of any pulmonary symptoms suggestive of TB by a physician or trained health care provider to determine if symptoms of TB disease have developed.
- *Reassessment Following Exposure or Change in Health Status.* Workers who experience exposure to an individual with suspect or confirmed infectious TB, for whom infection control precautions have not been taken, and employees who develop symptoms of TB disease shall be reassessed and treated in accordance with the CDC guidelines.

3. Base management of infected employees

Protocol for New Converters. Conversion to a positive TB skin test shall be followed up as soon as possible, by appropriate physical, laboratory, and radiographic evaluations to determine whether the employee has infectious TB disease and to begin preventive therapy if deemed appropriate.

Work Restrictions for Infectious Employees

- Employees receiving preventive treatment for latent TB infection should not be restricted from their usual work activities.
- Workers with latent TB infection who cannot take or who do not accept or complete a full course of preventive therapy should not be excluded from the workplace. These employees should be counseled about the risk for developing active TB and instructed regularly to seek prompt evaluation if signs or symptoms develop that could be caused by TB.
- Health care workers with pulmonary or laryngeal TB, while they are infectious, should be excluded from the workplace until they are noninfectious.
- Before returning to work, a worker who has had a bout of infectious TB should have documentation from the health care provider that the worker is receiving adequate therapy, the cough has resolved, and the worker has had three consecutive negative sputum smears collected on different days. After returning to work and resuming duties, while the worker remains on anti-TB therapy, facility staff should receive periodic documentation from the worker's health care provider that the worker is being maintained on effective drug therapy for the recommended time period and that the sputum acid-fast bacilli (AFB) smears continue to be negative.
- Workers with active laryngeal or pulmonary TB who discontinue treatment before they are cured should be evaluated promptly for infectiousness. If the evaluation determines that they are still infectious, they should be excluded from the workplace until treatment has been resumed, an adequate response to therapy has been documented, and three more consecutive sputum AFB smears collected on different days are negative.

4. Worker education and training

Training and information to ensure adequate employee knowledge of TB shall be provided to all current employees and to new workers upon hiring and be repeated as needed, or annually. Training should include:

- The basic concepts of TB transmission, pathogenesis, and diagnosis, including information concerning the difference between *latent* TB infection and *active* TB disease, the signs and symptoms of TB, and the possibility of *reinfection.*
- The potential for occupational exposure to persons who have infectious TB in the health care facility, including information concerning the prevalence of TB in the community and facility, the ability of the facility to isolate patients properly who have active TB, and situations with increased risk for exposure to TB.
- The principles and practices of infection control that reduce the risk for transmission of TB, including information concerning the hierarchy of TB infection-control measures and the written policies and procedures of the facility. Site-specific control measures should be provided to workers working in areas that require control measures in addition to those of basic TB infection-control program.
- The purpose of PPD skin testing, the significance of a positive PPD test result, and the importance of participating in the skin test program.
- The principles of preventive therapy for latent TB infection. These principles include the indications, use, effectiveness, and the potential adverse effects of the drugs.
- The worker's responsibility to seek prompt medical evaluation if a PPD test conversion occurs or if symptoms develop that could be caused by TB. Medical evaluation will enable workers who have TB to receive appropriate therapy and will help to prevent transmission of TB to residents and other workers.
- The principles of drug therapy for active TB.
- The importance of notifying the facility if the worker is diagnosed with active TB so that contact investigation procedures can be initiated.
- The responsibilities of the facility to maintain the confidentiality of the worker while ensuring that the worker who has TB receives appropriate therapy and is noninfectious before returning to duty.
- The higher risks associated with TB infection in persons who have HIV infection or other causes of severely impaired cell-mediated immunity, including the more frequent and rapid development of clinical TB after infection with TB; the differences in the clinical presentation of disease; and the high mortality rate associated with Multidrug-resistant TB in such persons.
- The potential development of cutaneous anergy as immune function (as measured by CD4 + T-lymphocyte counts) declines.
- Information regarding the efficacy and safety of BCG vaccination and the principles of PPD screening among BCG recipients.
- The facility's policy on voluntary work reassignment options for immunocompromised employees.
- Workers shall be trained to recognize and report to a designated person any resident with symptoms suggestive of infectious TB and instructed on the postexposure protocols to be followed in the event of an exposure incident.

5. Engineering controls

The use of each control measure must be based on its ability to abate the hazard. Individuals with *suspected* or confirmed infectious TB disease must be placed in a respiratory AFB isolation room. AFB isolation refers to a *negative-pressure* room or an area that *exhausts room air directly outside* or through HEPA filters if recirculation is unavoidable.

- Isolation rooms in use by individuals with suspected or confirmed infectious TB disease shall be kept under negative pressure to induce airflow into the room from all surrounding areas (e.g., corridors, ceiling plenums, plumbing chases, etc.)
- *Directional airflow*—General ventilation system should be designed and balanced so that airflows from less contaminated (i.e., cleaner) to more contaminated (less clean) areas to prevent the spread of contaminants to other areas.
- *Negative pressure* is attained by exhausting air from an area at a higher rate than air is being supplied. The level of negative pressure necessary to achieve the desired airflow will depend on the physical configuration of the ventilation system and area, including the airflow path and flow openings, and should be determined on an individual basis by an experienced ventilation engineer.
- Fixed room air recirculation systems, which recirculate the air in an entire room, may be designed to achieve negative pressure by discharging air outside the room.
- Some portable room air recirculation units are designed to discharge air to the outside to achieve negative pressure. Air cleaners that can accomplish this must be designed specifically for this purpose.
- A small centrifugal blower (exhaust fan) can be used to exhaust air to the outside through a window or outside wall. This approach maybe used as an interim measure to achieve negative pressure, but it provides no fresh air and suboptimal dilution.
- The employer must assure that AFB isolation rooms are maintained under negative pressure. At a minimum, the employer must use nonirritating smoke trails or some other indicator to demonstrate that direction of airflow is from the corridor into the isolation/-treatment room *with the door closed.*
- Air exhausted from AFB isolation rooms must be safely exhausted directly outside and not recirculated into the general ventilation system.
- Opening and closing of doors in an isolation room that is not equipped with an anteroom compromises the ability to maintain negative pressure in the room. For these rooms, the employer should utilize a combination of controls and practices to minimize spillage of contaminated air into the corridor. Recognized controls and practices include, but are not limited to, minimizing entry to the room, adjusting the hydraulic closer to slow the door movement and reduce displacement effects, adjusting doors to swing into the room where fire codes permit, avoiding placement of room exhaust intake near the door, etc.

Note: A HEPA section has been excluded from this document as it is rather detailed, inconclusive, and unlikely to be practical for many facilities.

Respiratory Protection

"Respirators shall be provided by the employer when such equipment is necessary to protect the health of the employee. The employer shall provide the respirators that are applicable and suitable for the purpose intended. The employer shall be responsible for the establishment and maintenance of a respiratory protective program which shall include the requirement outlined in paragraph (b)."

Requirements for a minimally acceptable program:

- Type 95 Respirators
- Fit test affected/assigned employees for respirators
- Provide at least three sizes of respirators to enhance fit
- Fit checks are performed by employees each time one is donned

Employees are to wear HEPA or certified respirators when:

- Entering a room housing individual with suspected or confirmed infectious TB.
- Emergency-medical response personnel or others transport, in a closed vehicle, an individual with suspected or confirmed infectious TB.
- Workers are present during the performance of high-hazard procedures on individuals who have suspected or confirmed infectious TB.

If a facility chooses to use disposable respirators as part of its respiratory protection program, their reuse by the same health care worker is permitted as long as the respirator maintains its structural and functional integrity and the filter material is not physically damaged or soiled. The facility must address the circumstances in which a disposable respirator will be considered to be contaminated and not available for reuse.

Note: When respiratory protection, including disposable respirators, is required, a complete respiratory protection program must be in place in accordance with the Respiratory Protection Standard. This includes:

- Written Respiratory Protection Program
- Medical Clearance to Wear a Respirator
- Fit Testing
- Annual Training
- Maintenance and Repair

Access to Medical and Exposure Records

CFR 1910.20. This provides the employee with the right to access his or her medical and exposure record. A record includes accident reports, MSDSs, physical examination reports, any other written reports that included the individual in their analysis. Employer must notify employees of this right. The employer must furnish requested information to the employee within 15 days of request at no charge to the employee. A representative of the employee may also obtain the same information with written permission from the employee.

- A record concerning employee exposure to TB is an employee exposure record within the meaning of this Standard.
- A record of TB skin test results and medical evaluations and treatment are employee medical records within the meaning of this Standard. Where known, the workers exposure record should contain a notation of the type of TB to which the employee was exposed (e.g., multidrug-resistant TB).

Accident Prevention Signs and Tags

CFR 1910.145. A warning shall be posted outside the respiratory isolation room. This Standard requires that a signal word (STOP, HALT, or NO ADMITTANCE) or biological hazard symbol be presented as well as a major message ("special respiratory isolation," or "respiratory isolation"), and a description of the necessary precautions (e.g., "Respirators must be donned before entering, respiratory isolation room or a sign must be posted referring visitors and employees to the nursing station for instruction"). The warning sign must state specifically the precautions required to interact with the residents. Indicators on residents' records, printed in language or symbols easily recognized by employees are additional methods to achieve this purpose.

- Employer shall also use biological hazard tags on air transport components (fans, ducts, filters) that identify TB hazards to employees associated with working on air systems that transport contaminated air.
- Biological hazard warning signs shall be used to signify the actual or potential presence of a biohazard and to identify equipment, containers, rooms, materials, which contain or are contaminated with viable hazardous agents.

OSHA 200 Log

- CFR 1904. For OSHA Form 200 record-keeping purposes, both TB infections (positive TB skin test) and TB disease are recordable in long-term care facilities. A positive skin test for TB, even on initial or baseline testing (except for preassignment or preemployment screening) is recordable on the OSHA 200 Log because there is a presumption of work-relatedness in this setting unless there is clear documentation that an outside exposure occurred.
- A positive TB skin test provided within 2 weeks of employment does not have to be recorded on the OSHA 200 Log. However, the initial test must be performed prior to any potential workplace exposure within the initial 2 weeks of employment.
- If the employee's TB infection was entered on the OSHA 200 Log, progresses to TB disease during the 5-year maintenance period, the original entry for the infection shall be updated to reflect the new information. Because it is difficult to determine if TB disease resulted from the source indicated by the skin test conversion or from subsequent exposures, only one case should be entered to avoid double counting.

Summary

Most long-term care facilities would be assessed as having minimal risk, or low risk. Almost all facilities have basic policies of moving any suspect individual to a hospital until cleared for return. However, a well-construed policy is called for to give management staff detailed guidelines on how to assess residents, while reassuring staff that the correct measures are indeed being taken.

Many facilities may also shrug off the need for any serious plan because the underlying policy is to remove the resident. However, all too often, suspect cases are dubious. There will be times when the decision whether or not to send a resident to the hospital is difficult to make. Sometimes, the decision is that the resident is not suspect, but is closely observed and is suspect to employees. Problems arise with employees filing complaints with OSHA when they feel that management has declined to acknowledge that a case is suspect, or is unable to handle a suspect TB case. There may be delays in the arrival of transporters to move the resident. There may be a number of things that can delay or impede the speedy removal of the suspect individual.

For these and other reasons (e.g., as a fail-safe, and affording a measure of confidence and capability to cope with changing circumstances), it is reasonable for a long-term care facility to have a meaningful TB protection program that addresses the shortfalls of the CDC and OSHA guides as discussed, affording it the luxury of control and conservatism should there be any question of suspicious TB symptoms. In particular, the following should be addressed:

Identify the most appropriate room for respiratory isolation based on the ventilation pattern of the facility. The ventilation of this room must be under negative pressure and the air must not recirculate throughout the rest of the facility (seal off recirculating exhaust vent from central air system, install exhaust fan in window). These are temporary measures for the time that a resident may need to remain on site before decisions, permissions, and other arrangements are made and transport is available.

Have a working Respiratory Protection Program. Include several employees from all shifts of all departments. In the event that a suspect resident is present and requires the services provided by nursing, housekeeping, or maintenance, these employees will be appropriately trained and protected to be able to attend to the needs and problems.

Attachments

- CDC Definition of Five Levels of Risk
- Checklist of a Complete TB Control Program

CDC DEFINITION OF FIVE LEVELS OF RISK

Minimal Risk—This applies only to an entire facility: Facility does not admit TB patients to inpatient or outpatient areas. Facility is located in a county or community in which no TB cases have been reported during the previous year.

Very Low Risk—This generally applies to an entire facility, except outpatient areas, which will fall into low-, intermediate-, or high-risk categories: Facility does not admit TB patient to inpatient areas but may initially assess, evaluate, or manage TB patients in outpatient areas (i.e., ambulatory care or emergency department). Patients who may have active TB and need inpatient care are promptly referred to collaborating facility. If patients are admitted to inpatient areas, this may still be appropriate for a limited number of other areas (i.e., administrative) or occupational groups that have only a very remote possibility of exposure to TB. May be appropriate for outpatient facilities that do not perform initial assessment for TB, but that screen patients for active TB as part of a limited medical screening before specialty care (i.e., dental setting).

Low Risk—PPD test conversion rate is less than the rate in areas or groups without occupational exposure to TB or previous rates in the same area. No clusters of PPD conversions (cluster means no more than two PPD conversions over a 3-month period and epidemiological evidence suggests nosocomial transmission). No evidence of person-to-person conversion. Fewer than six infectious patients/year.

Intermediate Risk—Same factors as low risk, except there are six or more infectious patients/year.

High Risk—Any one of these elements is enough: The PPD test conversion rate is significantly greater than that for control areas or groups (i.e., where occupational exposure is unlikely) or than previous conversion rates for the same area or group. There is a cluster of PPD test conversions, and epidemiological evidence suggests nosocomial transmission (see note in low risk). There is evidence of person-to-person transmission of TB. High-risk cough-inducing procedures are performed, such as sputum induction, aerosol treatment, endotracheal intubation and suctioning, and brochosocopy. Other procedures that may generate infectious aerosols are performed, such as irrigating TB abscesses or homogenizing or lyophilizing tissues.

CHECKLIST OF A COMPLETE TB CONTROL PROGRAM

Assignment of Responsibility
?_____ Identify person(s) with direct responsibility.

Risk Assessment and Periodic Reassessments
?_____ Obtain community TB profile.
?_____ Evaluate patient TB data.
?_____ Evaluate skin test conversions among health care workers.
?_____ Rule out person-to-person transmission.

Written TB Infection Control
?_____ Select initial risk protocol.
?_____ Develop written TB infection control protocols.

Repeated Risk Assessment at Appropriate Intervals
?_____ Review community, patient, and health care worker data.
?_____ Observe health care worker infection practices.
?_____ Evaluate maintenance engineering controls.

Identification, Evaluation, and Treatment of Patients Who Have TB
?_____ Create an index-of-suspicion checklist to screen patients on initial encounters in the emergency room or ambulatory care, and before or at the time of admissions.
?_____ Perform the most rapid laboratory tests on patients with suggestive signs of TB.
?_____ Promptly initiate treatment.

Managing Possibly Infectious Outpatients
?_____ Promptly initiate TB precautions.
?_____ Place in separate waiting rooms or isolation rooms.
?_____ Give surgical mask and tissues to patient, and instruct how to use them.

Managing Possibly Infectious Inpatients
?_____ Promptly isolate.
?_____ Monitor response to treatment.
?_____ Follow criteria for discontinuing isolation.

Engineering Design
?_____ Collaborate with ventilation expert.
?_____ Use single-pass air system, or HEPA filtration for air circulation in infectious TB patient care areas.
?_____ Consider supplemental measures such as UVGI.
?_____ Determine the number of isolation rooms needed, based on risk assessment and in consultation with regional public health department plans. (Existing isolation rooms must achieve a minimum of six ACH, air changes per hour, new or renovated rooms must achieve a minimum of 12 ACH.)
?_____ Regularly monitor and maintain all ventilation controls.
?_____ Perform daily tests for isolation rooms in use.
?_____ Exhaust TB isolation room air and local exhaust device air to the outside, or if absolutely impossible recirculate after HEPA filtration.

Respiratory Protection
?_____ Purchase masks meeting NIOSH (Natural Institute of Occupational Health and Safety) criteria.
Health care workers must wear masks when:
?_____ Entering TB isolation rooms.
?_____ Performing cough-inducing or aerosol-generating procedures on persons with confirmed or suspected infectious TB.
?_____ Performing autopsies on bodies of such persons.
?_____ Transporting such individuals.

?_____ Have a written respiratory protection program in place.

Health Care Worker TB Training and Education
?_____ Require for all health care workers.

?_____ Include epidemiology of TB in facility.
?_____ Emphasize pathogenesis and occupational risk.
?_____ Describe work practices to lessen exposure.

Health Care Worker Counseling and Screening
?_____ Counsel all health care workers regarding TB infection and the increased risk to immuno-compromised individuals.
?_____ Perform preemployment skin testing and then at regular intervals, at least anually and twice a year for high-risk health care workers.
?_____ Evaluate symptomatic health care workers for active TB.
?_____ Evaluate any health care worker skin test conversion or possible nosocomial transmission.
?_____ Coordinate efforts with local public health departments.

G. IMPORTANT TB RESPIRATOR NOTES

Read this before purchasing any respirators: Sometimes facilities invest in equipment that is not suitable to its needs. The following will help to determine what a facility needs, desires, and the specifications necessary to fulfill them.

NIOSH-Approved Respirators

NIOSH is the agency responsible for setting guidelines for approving respirators. Recently, it has drastically revised its guidelines, definitions, and standards for respirators. In an unusually compromising move, NIOSH has approved a new breed of approved masks for TB use. It devised a new system of categorizing respirators, and has approved the *N95 masks* for *TB isolation* room use. The N95 masks are a departure from the High-Efficiency Particular Air (HEPA) filters that were prescribed for TB based on the previous standard. HEPAs provide 99% removal efficiency that is optimal for filtering out all sizes of TB bacillus. Also, any and such filters on a normal respirator allow users to "user seal check" proper respirator-to-face seal before each use, as good and mandated protocol. In fit checking, the wearers form a seal over the filter pieces (traditionally with hands over the filter canisters placed into the respirator) and check for leaks by sucking air in and blowing air out to see if the mask collapses on their face, or lifts evenly off their face, respectively, to ensure proper face-to-mask seal.

The N95 masks provide a 95% removal efficiency, which is adequate, but not optimal for removing TB bacillus. The N95 "mask," not a respirator per se is difficult and, in most cases, impossible to "user seal check" properly. Manufacturer's instructions usually consist of asking users to cup their hands over the surface of the mask to create the seal as discussed above, which is not very practical. This N95 approval was granted to assuage the complaints of hospitals, which found the use of traditional respirators too costly, physically overbearing, and awkward to provide patient care. It is uncharacteristic of NIOSH to forgo stricter environmental health and hygiene standards and concede to economic considerations.

N95 Masks

In strict terms of compliance, the N95 masks will suffice for TB protection. They are rather inexpensive ($1 to $3 each). They are designed for single use only. Users discard them immediately after exiting the isolation room. They are difficult, if not impossible, to user seal check properly. Nonetheless, with NIOSH approval, the N95 mask will do.

Alternative—Disposable HEPA Respirators

The disposable HEPA mask/respirators or N99 provide a higher level of protection (99% vs. 95% removal efficiency, which removes smaller TB particles) and are of a more substantial design; some may be user seal checked. Many occupational health and safety professionals prefer the N99 over the N95 masks. Disposable HEPA masks/respirators that are available with fit cups are even better.

These allow for proper user seal checking before each use. Although they are designed to be one-use, disposable respirators, long-term care use would be so minimal in terms of length and severity of potential exposure that it could most likely be used for one incident, for as long as it takes to be resolved, one day, or several days, depending on length, severity, and potential of exposure.

When shopping for HEPA or N99 disposable respirators, try to find those that have fit cups. The Uvex brand does not exist anymore. It had made the disposable HEPA respirators with fit cups. These respirators with fit cups are now made by Better Breathing. 3M has disposable HEPA respirators, but they do not have fit cups to go with them. A 3M sales representative claimed that there would be liability involved in providing fit cups that have not been quantifiably proved to provide the actual seal to be able to execute a proper user seal check. Other manufacturers are willing to provide them as an assistive tool to help users gain a better fit, although it may not have been quantified in a laboratory. As a real-life practical user on the front line, all the help one can get in trying to gain a better fit, albeit not 100% guaranteed, is greatly appreciated.

In ordering these respirators, make sure that they come with fit cups, and, in fit checking, make sure to have different sizes available. It is actually best to have the small, medium, and large sizes of two different types or brands of these respirators to ensure optimal fitting during fit testing.

H.   SAMPLE PROGRAM: WRITTEN RESPIRATORY PROTECTION (FOR LONG-TERM CARE AS PART OF THE TB INFECTION CONTROL PROGRAM)

Introduction

OSHA requires a written respiratory protection program along with annual employee training, medical clearance, fit testing, and equipment maintenance for the use of any respirator, even if only one is used. The following sample program is designed specifically for long-term care to comply with this Respiratory Protection Standard, within the scope of an Infection Control Plan concerning TB. Thus, if you intend to use other types of respirators for other uses, i.e., for spray painting, you may add it into this program. However, the author suggests that you contact her office to ensure that all the necessary details appropriate to its addition will be included, since no guidance will be given herein for other respirators.

Section 1—Personnel and Fit Test Equipment

A *Program Coordinator* shall be appointed. This person shall be qualified by experience and appropriate educational background. All responsibilities and duties named in this plan shall fall ultimately under such person to act on or to delegate. The coordinator should be knowledgeable about the following: principles of respiratory protection, equipment technologies, applicable rules and regulations, scientific literature on toxicology, exposure limits, and substance specific aspects of respiratory protection as they relate to Tuberculosis, as well as appropriate administrative and technical procedures and protocols. This coordinator shall also keep abreast of the current trends and activities of the same.

The program coordinator shall be authorized and supported by the Corporate Respiratory Protection Program. This person shall be furnished with and be responsible for maintaining an adequate record-keeping system, and accessible licensed medical consultant and fit testing equipment.

Fit testing equipment should include a well-vented area where tests can be performed properly. Smoke tubes or banana oil or saccharin, an air bulb or nebulizer, disinfectant, sample respirators of varying types and sizes along with record-keeping forms should be available.

Section 2—Medical Surveillance

An *Initial Medical Clearance* must be obtained from a physician or other licensed health care professional (individual whose license, registration, or certification allows him or her to provide medical evaluation services) *before* any individual can be fit-tested for a respirator. This medical consultation shall determine what health and physical conditions are pertinent, i.e., an individual has any complications that may restrict respirator use. Cardiovascular, pulmonary, and allergy conditions should be reviewed.

Medical surveillance should be conducted periodically. This should include preemployment, periodic examinations to reaffirm the ability of the individuals to wear the equipment issued and to monitor for any deteriorating health effects that may be due to exposures, and at termination of employment. During these consultations, a baseline status of general health should also be noted. Annual review of medical status is no longer required unless physiological or physical change is observed in the employee; the physician or other licensed health care professional, supervisor, or program administrator informs the employee that an employee needs to be reevaluated; information from the respirator program, including observations made during fit testing and program evaluation indicates a need for change; or a change occurs in the workplace conditions that may substantially increase the physiological burden on an employee. (The more stringent and elaborate respiratory protection regulations for specific substances, i.e., lead, formaldehyde, etc., must be followed in the event that these substances are used on site and thus those special standards would apply.)

Section 3—Hazard Evaluation

As this written respiratory protection program is developed and implemented in conjunction with a TB Infection Control Plan, the hazard evaluation should be conducted, evaluated, and documented as part of that Infection Control Plan. The use of respirators in this context is basically for two purposes:

As good basic industrial hygiene, added measure of protection, or extra precaution in the event of a suspiciously "suspect" TB case, or when a case has not yet been diagnosed as TB, or not yet been diagnosed as "suspect," but the facility would prefer to resume care and not discharge the resident to a hospital for observations, while minimizing potential exposure to staff.

To provide a facility with the legal and practical ability to care for a resident who has been diagnosed with active TB, or diagnosed as "suspect" TB for the length of their diagnosis, or the beginning, or the end of their diagnosis, without ever having to discharge to a hospital.

In such cases, the hazard is specifically known to be tuberculosis. This sample policy is designed specifically for this use, and no other. [Again, if you plan to include other uses for respirators in this plan, please consult the author to determine what additional components are required to comply.] The hazard will be confined to the isolation room.

Isolation room

The isolation room should be properly maintained for:

— Adequate air exchange, at minimum, six air changes per hour (refer to Section III.E for a worksheet to determine air changes in any given room);
— No air recirculation into other parts of the facility, or air leakage (duct tape/seal);
— Ventilation to the outside (small window fan or bathroom vent if adequate);
— Sequestered outside ventilation area (prevent inadvertent exposure);
— Proper labeling on doors ("Authorized Personnel Only"/"Door Must Remain Closed At All Times");
— Consistently monitored for airflow pattern and volume;
— Respirator use will be designated to the specially selected and trained staff that will service the resident in the isolation room;
— Isolation rooms should be under negative pressure.

As of the date of this writing, OSHA still has no Standard on TB. However, OSHA expectations generally are that facilities follow the NIOSH and CDC guidelines on TB.

Industrial hygiene monitoring

There are no set limits, i.e., permissible exposure limits (PEL), or ceiling limits, or short-term exposure limits (STEL), or time-weighted average (TWA) limits for the level of allowable exposure to TB bacilli. There probably will not be one, or at least any time too soon, because there is controversy among the scientific community over what concentrations and what types of exposures will lead to infection. *Thus, requirements for biological monitoring for exposure levels of bacilli do not seem likely in the foreseeable future.* The only monitoring involved is for airflow around and in the isolation room.

Hierarchy of controls: engineering → administrative → personal protective equipment

In dealing with any given hazard, it is the general rule of health and safety strategies, to engineer a hazard out of risk first (i.e., machine guard, ventilation). Then, if that is not possible, or still not adequate, administrative means should be next (i.e., controlling staff and shifts to spread and dilute exposure, or in this case, concentrate exposure for better control, developing isolation room rules, and enforcing them, etc.). The least effective and, thus, the last strategy of controlling hazards is personal protective equipment.

In the case of TB exposure control, all three methods are used. Initially, exposure to a hazard is engineered to a minimum by way of a working isolation room. Second, administrative measures are taken, including the writing of this plan, determination of when to use the isolation room, the selection and training of employees. Last, respirators are used.

Section 4—Fit Testing

Fit testing should be performed after the hazard evaluation process is completed. Comprehensive training on appropriate respirator and respirator accessories should accompany training for proper respirator care and fit during the fit test briefing. An important correlation should be emphasized between the type of equipment chosen and the type of hazard found. Thus, staff should understand that the respirators issued for TB exposure control should not be interchanged with other types of respirators, or used for other purposes. Fit tests are now required on an annual basis.

Proper fit, care, and use of respirators will only be beneficial if the appropriate respirators and accessories are also used. It may prove to be useless or sometimes even harmful (giving employees a false sense of security) if the wrong type of equipment is properly donned.

Care should be taken that respirators are not issued or used until the employee has attained medical clearance, has been properly fit tested, and has been trained to use, maintain, and check the equipment. The level of protection and the type of respirator used must match the hazard of the work involved, as previously quantified or qualified in the hazard evaluation.

The simplest fit testing method is the qualitative fit test using smoke tubes, banana oil, or saccharin. The employee tested should be given a choice of several respirators of varying sizes and models if necessary. The employees chooses the one with the closest estimated fit. The Fit Test Agent shall demonstrate how to don a respirator properly.

The employee should be given at least 10 minutes to adapt to the respirator. Meanwhile, the agent can explain and show how to check for leaks and a good facial seal. For respirator use with fit cups available, exhale with fit cup covering the surface of the respirator to feel a lift off the face, and inhale with fit cup on the surface of the same to feel the suction onto the face. If one is using an N95 respirator that does not have fit cups, place hands over the surface area of the respirator to create a seal and test for suction and lift off, respectively.

The employee should be instructed to move his or her head up and down and side to side while talking to adjust for comfort and seal. If at any point there is unreasonable discomfort or inadequate fitting or facial seal, another respirator size or model should be tried.

In a well-ventilated area, instruct the employee to close his or her eyes, and count, or repeat the alphabet throughout this entire process. Closed eyes are most important when using irritant smoke. Have the wearer turn the face to the right, and spray the smoke or mist at the left cheek, turn to the left while spraying the right cheek, head turned up and spray under the chin, and head bent down while spraying over the top of the head. The wearer should remain in the smoke or mist cloud for several minutes while continuing to speak. Eyeglasses should not interfere with the fit of the respirator and should sit on top of the bridge of the nose, above the respirator seal area.

If the wearer experiences any coughing or irritation at any time during these tests, he or she should be led away from the smoked area immediately and should remove the respirator. If the wearer can detect any sweet odors from the mist, then the area should be ventilated. In either case, there has been a leak and the seal is not effective. Discontinue the test and begin again with a different size or type of respirator with a better seal.

Individuals with facial hair, including beards or shadow growth, should not be fit tested. This would interfere with the facial seal of the respirator. The employee then cannot obtain or use a respirator, or work in the area of concern. Employees should be made aware of this facial hair policy and it should be enforced continually.

Section 5—Record Keeping

The Fit Testing Agent must maintain a clear record of the fit-testing procedures. The date of testing, the employee name, job function, the name and signature of the person conducting the test, the size of the respirator chosen. These records should also reflect the type of respirator issued, and that fit testing is done annually.

Section 6—Equipment Selection

For the purpose of this Plan, a HEPA (High Efficiency Particulate Air) or the N99 equivalent or a N95 rated respirator according to the new NIOSH respirator rating scheme would be appropriate for TB isolation room use. Note that respirators designed for this particular use in health care settings do not tend to have the exhale/inhale valves of ordinary respirators. Fit checking is conducted with a fit test cup for better-designed HEPA respirators. For respirators of inferior design, the use of one's hands over the surface area of the respirator, to create the seal to check for fit, in the absence of a fit cup, is recommended. This has been approved by NIOSH, but is not as sound a design as those respirators that come with a fit test cup. The N95s, however, are much more economical.

For some examples of respirators, review any occupational health and safety product catalog. For example, Lab Safety Supply at 800-543-9910 carries N95 respirators for as low as $13.40 a box (box containing 20) for the RACAL 95 (order number 7E-32000), and up. HEPAs can start at about $145.40 per box of 20. Remember when ordering HEPAs, the fit test cups are extra ($4.25 each). [Section III.G.]

Section 7—Employee Training

Respiratory Protection Program Training is required for all employees who are covered by this respiratory protection program. Training should be given to new incoming employees and existing employees with new duties that require the use of respirators; continuing education and yearly refresher training is required for all other affected employees. Because of the nature of the specificity of this hazard, this training should be incorporated in the TB annual training. Specific training on proper respirator use may be limited to those who have been selected.

The topics covered in training should include introduction to the Respiratory Protection Program, TB exposure, limits of respirators, facial hair policy, etc. Employees should be trained to inspect their respirators before each use. They should check for cracks and signs of deterioration. They should be trained on proper storage and maintenance. Respirators should not be stored in a way that they can be crushed or disfigured. Unlike ordinary respirators, they should not be shared, because they cannot be disinfected without adversely affecting the material of the respirator.

Employees should be trained to check for fit before and after each use. Employees with respirators that have fit cups should be issued one each. Fit cups should be properly placed over the respirator to check for fit as defined in the Test Fitting Section of this plan. Test fitting should also be carried out by employees using respirators without fit cups, using their hands instead.

Employees should also be made aware of the need to reevaluate the respirator fit if they experience major weight, dental, or facial changes, as these may affect the seal of the respirator. They should also be made aware of when they may need to exchange a respirator for a new one.

Respirator Life. The general rule of thumb for respirator life in hospitals is about one per shift. Because of this high rate of change, hospitals prefer the N95 respirators, over the HEPA or N99, because, although inferior in design, they are much cheaper. However, hospital respirator use is different from that of long-term care use. The exposure is higher, and probably longer, in hospitals.

Thus, respirator life span in long-term care could be extended depending on the length of exposure (time actually spent inside the isolation room), which is also dependent on the length of the resident's stay in the isolation room and the potential infectiousness of the resident. The scenario can dictate the necessity for strict filter-change frequency, for example, if the resident was being kept as a diagnosed active TB case, or a "suspect" case, or merely kept in isolation for good measure awaiting the final outcome of diagnosis, and the symptoms displayed by the resident (from little to no symptoms to excessive coughing, spitting, aspirating, etc.).

In most cases, which are mild and highly conservative, using HEPAs, or N99s, for the length of the resident's stay in isolation, even for a period of several days, would still probably be prudent if exposures are very low-risk and brief. If exposure is strongly suggested, and length and severity of exposure is also serious, then consideration for more frequent filter changes should be implemented.

IV.   WORKPLACE VIOLENCE PREVENTION

A.   GOAL

Employers must manage the potential for workplace violence hazards as much as possible within reason, depending upon their industry and individual circumstances.

B.   SUMMARY

Although there is no OSHA standard on workplace violence, it is a *recognized* hazard for most industries. Few businesses, if any, are insulated from the potential risks. In addition, there are legal ramifications beyond OSHA when employers allow employees, or other persons within their domain, to act as perpetrators and use company phones, vehicles, property, and other equipment to commit workplace violence.

Long-term care facilities need to manage the risks inherent in operating a 24-hour facility that maintains a certain amount of semipublic traffic during daylight hours with parking lots and possibly some secluded areas indoors and outdoors, along with providing care for demented residents who may become combative at times. This is in addition to the normal risks involved in any employee population, which may contain at-risk individuals and individuals with personal domestic issues that may spill over into the workplace. The majority of workplace violence incidents involve domestic violence, in general industry. For long-term care, combative residents may be the leading cause depending on patient population and the strength of the facility's special care or dementia programming. Many other incidents are precipitated by conflict between employees. Some of these types of incidents actually become workers' compensation claims.

Annual training should include an overview of the facility's policy on workplace violence. Many facilities have experienced success in seeking the assistance of community domestic violence prevention or crisis centers to provide and/or supplement the training.

C. UPDATE

OSHA inspectors will ask to review facility policy on workplace violence prevention, including any program dealing with combative residents if the occurrence is discovered to be present during employee interviews and/or OSHA 200 Log entry reviews. They will also review annual employee training on the subject.

D. CONTENTS

Following is a sample workplace violence prevention management policy guide (Section IV.E) and a related guide on workplace violence prevention, risk reduction, policy on resident behavior management (Section IV.F).

E. SAMPLE POLICY GUIDE: WORKPLACE VIOLENCE PREVENTION MANAGEMENT

Introduction

The prevalence and type of workplace violence is very specific to geography, logistics of the building and its grounds, and the socioeconomics of the surrounding neighborhood. Therefore, this outline is designed to guide users through issues for consideration, rather than specific items to include in a policy.

Purpose

This guide outlines a facility's policy and establishes guidelines in managing and reducing the potential for workplace violence at the facility.

Objective

The objective is to ensure the highest standard of health and safety for all employees, residents, vendors, contractors, and the general public and to provide for the efficient and effective operation of the facility.

Definitions

The following terminology is defined as set forth by the Workplace Violence Prevention Research Institute, to establish and clarify terms and expectations.

Verbal harassment: Verbal threats toward persons or property; the use of vulgar or profane language toward others, disparaging or derogatory comments or slurs, offensive sexual flirtations and propositions, verbal intimidation, exaggerated criticism, and name calling.

Physical harassment: Any physical assault such as hitting, pushing, kicking, holding, impeding, or blocking the movement of another person.

Visual harassment: Derogatory or offensive posters, cartoons, publications, or drawings.

Appropriate Responses to Harassment or Threat

These should include the protocols staff are to follow when confronted with a threat or a harassing situation with a member of the public, visitor, contractor, or co-worker. A facility may want to distinguish expected appropriate behaviors based on the degree of threat or harassment. This applies to dealings with members of the public, visitors, family, vendors, contractors, consultants, and co-workers.

    Example: At any time an employee feels verbally or visually harassed by a visitor, vendor, or associate, the employee should excuse him or herself, refrain from arguing, retorting, or responding in kind, and notify the supervisor or a member of management.

    Example: At any time an employee *realizes* that he or she, or the people around the employee, are seriously threatened, or are in great danger (believes that serious and *probable* bodily harm exists, i.e., display of a weapon or very volatile behaviors), the employee should ask someone else as quickly as possible to notify supervisors and management, and call for help.

Protocols for Visitors/Vendors/Associates

1. Specify protocols for visitors, vendors, associates, or anyone else entering the building.
2. Specify protocols for after hours, when there is limited staff and no receptionist.
3. Specify protocols for a wandering stranger in the facility.

Example: All persons entering into the building must immediately verify to the receptionist, their identity, the person they will be meeting, and the purpose of their visit. They must sign in. The receptionist will confirm the information and direct the individual to the appropriate place, or to wait in the waiting area.

Example: Visitors, vendors, or associates who are asked to arrive at a time during off-hours, or when there may be no receptionist, should be previously instructed to sign in and proceed directly to the place of business and/or a specific individual to "check-in."

Example: Any visitors, vendors, or associates seen to be loitering, wandering, or lost, especially during off-hours, should be addressed immediately by available personnel, either personally, or by immediately contacting appropriate personnel. [Specify here, exactly what titles have this responsibility.] Personnel should politely ask if they can help the individual, verify the individual's identify and business, ask the individual to sign in, and direct the individual to the appropriate place and/or person. Personnel should exercise discretion whether or not they should accompany the individual until he or she has "checked in."

Prohibited Items on Property

The author suggests that all facilities adopt this policy:

Under no circumstances are the following items permitted on facility property, including parking areas: all types of firearms; switchblade knives and knives with a blade longer than 4 inches; dangerous chemicals; explosives including blasting caps; chains and other objects carried for the purpose of injuring or intimidating. This applies to *any* person on site at the facility including employees, visitors, contractors, etc.

Employees, visitors, and vendors alike should be discouraged from bringing and leaving any valuables on facility grounds without purpose(s) specific to their business or visit there. Some facilities provide for a check-in service for valuable items that were necessary to be brought in and stored for the day (i.e., gift items kept for the day for an event immediately after work, special vendor equipment requiring storage overnight, etc.).

Reporting of Incidents

As with the reporting of any and all incidents and accidents, employees should be urged to immediately report any and all incidents of harassment, threats, intimidation, or acts of violence to their supervisor, or another supervisor or the director if the problem is their supervisor. The supervisor must report the incident to the department head immediately, or as soon as practicable.

Investigation of Reported Cases

Selected members of management must review the facts of all reported cases of threats, harassment, intimidation, and acts of violence among employees and nonemployees to determine validity, degree of offense, and the proper response. This may range from interviews, eyewitness accounts, collection of evidence, and verification. Written statements are to be taken. A file or log should be initiated to track developments and provide a document trail for all activities of investigation, findings, and the response.

The investigation team should usually consist of the direct supervisor, department head, human resource, and any other persons who are connected with the business of the offending individual. This may include the social worker, or admissions person if the reported offender is a resident or family member, or the employee that is dealing with a reported offending vendor, etc. Of course, if the reported offending individual is, in fact one of these, that individual shall not serve on the investigation team.

The depth and immediacy of the investigation and the timeliness of its completion should be dictated by the degree and immediacy of the situation. However, the investigation and formal response should be completed and documented within 7 days.

The appropriate response can range from a short informal or formal meeting with the offending individual, explaining the facility's policy on conduct. Or, it may take the form of progressive discipline, to termination for employees. In the case of visitors, it can take the form of progressive warnings, to a ban from certain parts of the facility (e.g., restricted to the lobby) or the entire facility grounds. The same would apply in the case of vendors. The facility may deem it prudent to cease doing business with the vendor, or contact the management, inform it about the situation, and see if it will respond satisfactorily. In any case, depending on circumstances, it may even call for immediately contacting the police and an ambulance and pressing charges. Obviously, appropriate responses should be aligned with the severity and validity of the offense, the probability of risk, the resulting damage, if any. It is best to document how the investigation team arrived at their decision to respond to illustrate the judgment used.

Fair and reasonable alternative investigation procedures must be available in the event that it is a member of management who has been reported, particularly, higher and top management personnel.

Restraining Orders

Employees may be victims of domestic violence or stalkers. These issues may cross the boundaries from the personal realm and into the workplace as their perpetrators may follow them into the workplace or try to find them there. It is imperative that management be cognizant of such problems, and the impact they could have on its operations. Management should communicate its desire to be aware of such problems and its willingness to cooperate with employees who require restraining orders to keep such issues at bay. This provides a certain degree of comfort for the affected employee. Such proactive activities also provide for the facility's overall security.

[*Note:* Incidents of resident abuse perpetrated by visiting family members do occur. The management of such incidents is beyond the scope of this policy. Proper protocols in reporting and handling such incidents should include discussions with the social worker and admissions coordinator, and may require notification of DOH, and/or the ombudsman.]

Employees must notify management of any attempts to gain a restraining order, or of restraining orders that do exist, for spouses or stalkers. Such employees should supply the facility with a description of the individual (a photograph if possible), the history of violence or harassment in this relationship, whether the individual has access to firearms, weapons, or other potentially destructive and violent tools or materials, and as much other information as possible regarding potential times, days, or periods this individual may be more actively seeking to enter the facility. Management will ensure facility security, and selected management personnel will be made aware of and kept up-to-date on this information.

Former Employees

Policy should be set for the proper handling of uninvited returning former employees who are acquainted with the grounds and the personnel. Most former employees who wander into the facility uninvited and sometimes unannounced are entering on neutral, if not friendly, terms. However, if an employee was terminated on grounds of inappropriate conduct, or on poor terms, and was considered a security risk at the time, the uninvited return can also pose a significant risk. Policy must dictate how to differentiate these uninvited appearances, and how to handle the ones with potential risk.

Individuals who may pose a potential risk should be identified at the time of discharge. Appropriate personnel should be notified of the status of individuals as being *identified risk discharged employee* and of accompanying protocols to follow should the individual return. This could include immediately contacting the administrator or human resource director to speak to the individual, prevent any disruption, diffuse any developing hostility, and escort the individual out. If the individual refuses to leave and presents a continued risk, the police may have to be contacted to remove the individual. Repeated visits that dramatically threaten the safety and security of employees and residents may call for the pursuit of a restraining order.

Substance Abuse

References and highlights can be made here to the facility's policy on drug and alcohol use and abuse. It should be underscored that a great majority (77%) of acts of violence are perpetrated by individuals with substance abuse problems.

Criminal Activities

Any and all incidents of suspect criminal activity will be thoroughly investigated and referred to local authorities. Anyone (employee, visitor, contractor, etc.) found perpetrating a criminal act within facility grounds, including its parking lot, will be immediately removed. Employees and contractors will be terminated.

Locker Inspections

Routine unannounced locker inspections, with the employee present, may be instituted to promote good hygiene and housekeeping, and to deter the storage of stolen items, weapons, alcohol, and illicit drugs.

Bag Searches

Facilities may reserve the right to inspect bags that enter and leave the facility. Some facilities provide a checking service for employees bringing in large or expensive items that they must have with them on special occasions [see section on Prohibited Items]. Again, this provides for screening out weapons, drugs, alcohol, expensive items vulnerable to theft, etc. It also screens for stolen items leaving the premises.

Note: It is beyond the scope of this sample policy on workplace violence prevention to cover theft control issues, although it may be appropriate to make references to theft control policies at this point.

Riots and Civil Disobedience

Facilities should review their likelihood of being a target for hate or bias crimes, especially for facilities of a particular faith, vocation, or heritage, and a facility's general vulnerability during civil disobedience (looting and riots due to loss of power or civil unrest). Facilities should plan for these events, including an assessment of the adequacy of security on the grounds and procedures under such circumstances.

Security and Security Personnel

Is the security of the grounds, property, and people adequate? A nursing home strives for a relaxed, homelike atmosphere. However, it may be subject to heavy traffic at times. Without proper procedures, it can seem like a public thoroughfare. Part of the security question also depends on the specific layout of the facility, entrance area, and commonly trafficked areas. Can strangers, unfamiliar with the grounds, easily access other parts of the facility before passing areas of administration or can they immediately access large groups of residents?

Keep in mind that there is an entire population of highly vulnerable residents, along with a staff (a skeletal one for certain shifts), strongly dominated by female paraprofessionals and aides. If nothing else, a facility must consider the steps a receptionist, or the night shift, can take should an altercation occur.

Where necessary, security personnel with a certain degree of personal, professional, and physical strength may be considered. Such persons must be able to handle all types of visitors appropriately. These individuals must be able to warmly and courteously receive familial members; professionally receive and direct associates and vendors; provide assistance, direction, implement proper procedures; escort staff to parked cars or other remote areas in at-risk situations; perform bag searches; perform locker inspections with appropriate personnel; confront difficult or volatile individuals; and pursue and apprehend disruptive, violent individuals and intruders.

As mentioned above, where parking and parking lots pose a risk, staff should be able to request a security escort to their cars, or other remote, unsecured, at-risk areas.

Assessment and Abatement—Workplace Analysis and Follow-Through

An initial and periodic (at least annually) evaluation will be conducted on the grounds, and on security protocols, to assess the continued and improved effectiveness of this policy. This includes, but is not limited to, review of:

- Secluded and remote areas, solitary activities
- Sufficient and operational lighting, fencing, monitoring, locks
- Evening and night shifts
- Routes of access/egress, parking lots/areas, and safe access to them
- Employee screening protocols
- Termination protocols
- Employee training
- Effectiveness of reception area to screen and retain visitors properly
- Rapport with local police in sharing information
- Effectiveness in handling difficult and at-risk visitors, etc.

The following questions are taken from the section on security in Appendix D of this manual:

- Are security risks considered in the work site analysis?
- Is there adequate lighting in remote areas outside and inside the facility?
- Are TV monitors available to survey obscured areas?
- Do staff, especially "at-risk" staff (night and early shifts), receive training on handling potentially disruptive situations and victim-avoidance techniques?
- Is the parking lot secure?
- Are measures taken to minimize theft and robbery both within and immediately outside the facility?
- Are all persons entering and exiting the facility properly identified and documented, including their business at the facility?
- Is there a thorough criminal and drug abuse preemployment screening?
- Is the level of security at the facility compatible with the surrounding community's crime level and experience?
- If necessary, has a silent alarm system been considered or installed?
- If necessary, does the facility have a liaison, relationship, or communique with the local police department?

• Does the facility have adequate security personnel, especially during the night and evening shifts?

Training

Employees are trained during orientation and as annual security training sessions about workplace violence prevention policies, and proper protocols.

F. Sample Policy: Workplace Violence Prevention, Risk Reduction—Sample Policy on Resident Behavior Management (Corollary to Quality of Life, Special Care, and Residents' Rights)

Note: OSHA does not have a standard on workplace violence to date. It has published a guide for "Workplace Violence Prevention in Health Care." It is basic and includes material on hospitals and clinics that do not pertain to long-term care.

OSHA citations would occur under the General Duty Clause as there is no specific workplace violence standard. OSHA would argue that the cited facility has not adequately recognized and/or abated a workplace violence risk or hazard. This may be found in inadequacies in a facility's evaluation/assessment of its workplace violence risks, or its abatement methods (which includes employee training), and its follow-up, or, basically, its management of the issue.

This sample policy is a practical guide to formulating an in-house workplace violence prevention policy on the primary workplace violence issue for long-term care, employee injuries as a result of behavioral residents.

Purpose

This sample policy outlines a long-term care facility's policy and establishes guidelines and expectations on the prevention of aggressive, combative resident behavior, and, as a corollary, the promotion of quality of life for residents.

Objective

The objective is to ensure the highest standard of quality of life for residents, and health and safety for all employees, residents, visitors, vendors, and contractors, in an effort to provide for the efficient, effective operation of the facility.

Evaluation/Assessment

Preadmissions

Any indications of violent behaviors are to be noted from hospital reports, family interviews, medical and pharmacological history, and preadmissions screening interviews. Admissions, along with the Nursing Department and relevant departments and staff, must evaluate the ability of the long-term care facility to care for the potential resident properly and safely, given his or her medical, emotional, spiritual, and psychological circumstances and needs. If the resident record exhibits any risk of aggressive behavior, and the decision is made to admit, then appropriate steps must be taken to ensure that proper care can be provided for the individual. Staff are appropriately notified and trained to care for the individual.

Upon admission

At arrival, all residents are closely monitored for any signs of aggressiveness, combativeness, or antisocial behaviors. It should be noted what environments, activities, and behaviors by others may trigger the aggressive or antisocial behavior, and what environments, activities, and behaviors by others may prevent it and/or diffuse it. This is also done in conjunction with any medications or counseling that may be administered by the psychiatrist or psychologist.

Behaviors

Behaviors to be noted include but are not limited to those described in the "Mood and Behavior Patterns" of the MDS+ or the HCFAs Resident Assessment Instrument Manual. These include negative physical or verbal expressions of sadness or anxiety; distressed mood (miserable, blue, hopelessness, emptiness, tearfulness); repetitive questions; repetitive verbalizations (calling out for help); persistent anger with self or others; self-deprecation; unrealistic fears; repetitive anxious complaints; insomnia; apathy; crying; wandering; pacing; hand wringing; restlessness; fidgeting; reduced social interaction; resisting care (medications, injections, activities of daily living, assistance with eating, as opposed to exercising informed choice not to follow a course of care); verbally abusive (threatening, screaming, or cursing at other residents or staff), physically abusive (hitting, shoving, scratching, abusing sexually); socially inappropriate/disruptive behaviors (disrobing in public, excessive or inappropriate noises, screams, self-abusive acts, smearing/throwing food or feces, hoarding, rummaging through others' belongings, etc.).

14–Day psychological review

The frequency, severity, and persistence of the above-noted behaviors are assessed. The psychiatrist may indicate medications, behavior management, programming, and/or prescribe certain recreational therapies.

Hazard Abatement

Interventions and Care Planning

This is an ongoing process wherein care staff and medical professionals attempt to find and devise different interventions by trial and error to assuage disruptive behaviors and promote the resident's wellness and quality of life.

This can include but is not limited to changing sites, reality orientation, validation, different activities, diversions, music, physical activity, toys, merry walkers, memory lane, photo albums, etc. This is noted in care planning meetings and in the care plan that directs the specific care each individual resident should receive.

Communication

Care planning should involve the input of CNAs and orderlies providing direct patient care. Frontline staff provide feedback to the care planning group on new developments, observations, and experiences. As care plans are updated with their involvement, they will know immediately the reasons behind the decisions and directions of the plan.

Shift logs are kept and shared in order that developments, observations, and activities are recorded for the edification of the on-coming shift, for the sharing of information between shifts, and for the recording of observable trends.

There must be a good flow of communication, both formal (care planning) and informal, among direct patient care givers, family members, paraprofessionals, and the professional staff.

Dementia programs

There are special programs during the day wherein residents with dementia, or even different levels of residents with dementia, are grouped together, for certain activities, *with specially selected and trained staff*, geared toward their distinct needs. [Grouping such residents together, in and of itself, does not constitute a working dementia program.] Special care programs require sequestered and secluded environments with adequate space and design, along with the commitment of administration and the nursing leadership to implementing a truly special care program. [This involves changes in conventional and traditional attitudes and approaches toward resident relations and care, along with an understanding of value added employees.]

Staff selection and training is key in both dementia program(s) and special care unit(s). Staff must be specifically trained on disease processes, coping strategies, etc. Staffing levels can remain about the same, although more staff is better. Activities of daily living are executed in a more personalized and less institutionalized manner. These and recreational activities are geared toward the group's unique interests, intellectual ability, emotional sensitivity, concentration level, spiritual needs, physical ability, and emotional need to bond. Residents must be selected based on medical and personal background and history of behaviors to match the group, what the program(s) should offer, in concert with resident profiles and resident needs.

Such programs, if executed properly, provide an environment and atmosphere where relationships between care givers and residents are nurtured, and the individual needs of residents are truly attended to. This results in less agitated states and less disruptive, aggressive, or combative behaviors in residents, thus reducing the risk of injury to employees and other residents, while increasing quality of life for all.

Working dementia or special care programs result in an enriched quality of life for the residents who are having their particular needs better met; reduced risks of injury to employees; and increased employee job satisfaction; all through better resident management. Quality programs thus have a positive impact on the cost of workers' compensation premium by decreasing the probability and severity of claims arising from combative residents as a result of reduced risks, and increased job satisfaction. This also reduces turnover and absenteeism, and their associated costs in overtime pay, and new hires (orientation, start-up). It also serves to demonstrate strongly to OSHA that the facility is diligently managing what the agency considers to be the main "workplace violence" issue for this industry.

Special care units

Special care units are units where residents with dementia or different levels of dementia are grouped together on a wing or unit, and the dementia programs are held on that unit [whereas dementia programs are sometimes held in a central location and the attendees are brought in from different units].

Again, staff selection and training are key here, as well. Residents are also selected specifically for participation in these units. Special care units should be similar to an expanded dementia program where bathing, feeding, and all other activities take place under the auspices of special care, personalized regimen with an eye toward deinstitutionalizing the care, *and building relationships*.

Support groups

Support group(s) for employees should be formed and led by an appropriate social worker or similar individual. Support groups should meet once a week, and be made accessible to employees of all shifts. Employees can share experiences, vent, gain support, empathy, new ideas or strategies for coping, and regenerate. It should be driven by employee needs, and involvement and leadership. The group facilitator can also provide some structure and guidance. Provide members with the opportunity to choose their leader/facilitator and style of directing the group. Do not call it a support group. Use a more-positive name, "Employee Club," or "Employee Group." Employees have expressed reservations about the need to attend a "support group" because the term itself tends to have negative connotations of dysfunction, vulnerability, and neediness.

Policy training

At least annually, employees are to be made aware of and refreshed about the contents of this and the other related workplace violence prevention policies.

Coping strategies training

Training on intervention of dangerous or extremely difficult behaviors is rather unlike most other training provided in long-term care, or health care. With most clinical training in these settings, employers and employees can "get by" with just the "what to do" portion of an issue, and miss the "how to do it" and the extremely important and comprehensive "why." Unfortunately, this is often the case and health care may be more vulnerable as our medical model focuses so much on clinical techniques and often neglects or overlooks the patient's psychoemotional state and response.

As a result, many employees may be able to demonstrate almost perfect body mechanics and patient transfer techniques when asked. But few rarely practice it on a regular routine, "all the time, every time" basis, which is necessary for meaningful risk reduction. Meanwhile, the existing support system of cooperation, teamwork, and communication may not be aligned with this mandate, yielding less than optimal results.

In this way, much of the essence of coping strategies training would be all but lost, since it consists practically of 100% of the "how" and not "what." It is truly special training, as operators have begun to realize, since it involves "special care." The type of training needed cannot be accomplished in 30 minutes, or 1 hour, or 1 day for that matter. Nor can it be done in a typical teacher style, one-way monologue, or limited dialogue. The training must engender personal involvement from the employee. *Attendees must evolve through a process of empathy and understanding* of the disease, the patient, and their families. This requires time; in-services that include over 80% employee involvement; in-services that are almost always directed by employee experiences; confidential, secure, and relaxed settings that allow for exploration and self-reflection.

This very much requires an administration that truly understands these parameters and that will support the training needs of staff. It must also support the ideas, strategies, and efforts of those staff and units attempting to devise new and better methods of serving this client sector. This will include deinstitutionalizing the experience, designing for more flexibility and consideration for the human element, for both residents and staff, throughout operations wherever possible.

In terms of content, training should be reviewed at least annually, and preferably should include ongoing training on disease processes of dementia including Alzheimer's, Parkinson's, etc.; coping strategies with special care residents (including redirecting, refocusing, when to ignore behaviors, how to modify behaviors with positive reinforcement, how not to reinforce unwanted behaviors, etc.); how to communicate with residents with dementia; appropriate conduct for employees for different stages of dementia; appropriate activities for different stages of dementia; how to conduct activities of daily living with residents requiring special care; sensitivity training on the need to treat residents with dignity, respect, courtesy, and consideration; the importance of developing individual relationships with residents to promote emotional health and well-being and to reduce the likelihood of the development of antisocial, aggressive behaviors.

5   In-Service Records

I.   INTRODUCTION

A.   GOAL

Requirements for training records attempt to ensure that basic periodic safety training for specific hazards has been provided to employees.

B.   SUMMARY

Generally speaking, OSHA inspections will carefully review the training records for the following subjects in long-term care facilities:

- Bloodborne pathogens
- Fire extinguisher use (instruction on proper use, how to avoid misuse, and/or conservative use)
- Emergency evacuation (which must include evacuation routes, meetingplaces, and a system of employee census)
- Hazard communication
- Lockout/tagout
- Ergonomics
- Tuberculosis (TB)
- Workplace violence prevention
- Asbestos awareness (if employees work near or around asbestos or potentially asbestos-containing materials, i.e., floor or ceiling tiles, insulation, etc.)

Although OSHA does not have any specific standard on ergonomics, TB, or workplace violence prevention, inspectors will ask to review these since they are highly relevant to the industry's operations. Review of these records is considered to be within the scope of a general inspection of a long-term care facility.

In reviewing the training records, inspectors will look for the following:

- Training was provided *for these subjects* (training records that are not relevant to these subjects will not interest or impress the inspector in the least).
- Training was given during *orientation* and at least *annually* thereafter (for many facilities, the turnover rate is such that it is one and the same).
- Training covers all required elements within a topic.
- There was adequate *attendance* by appropriate employees.

If an inspector is conducting a "Records Inspection" particularly, 100% attendance will be required. If the records review is part of a larger overall inspection, then attendance will more likely be spot-checked as spurred on by results of employee interviews. However, it is not uncommon for inspectors to compare payroll with attendance to ensure adequate attendance.

OSHA inspectors search training documentation by:

- Subject
- Evidence of annual training

- Adequate class content
- Adequate attendance

Training records should be held for 3 years.

Respiratory protection training would apply if *any* respirators are used at the facility, even just *one* by a maintenance worker. (Any respirator use involves the need for a Respiratory Protection Program, see TB Chapter 4, Section III in this manual.)

Documentation of training in eyewash procedures and personal protective equipment training is not generally sought in the same manner as the above-mentioned subjects. However, it can become an issue when employee interviews reveal ignorance of the whereabouts and the proper use of personal protective equipment.

Job-specific safety training is also expected to be provided. Employees are expected to be trained in how to perform their tasks safely. However, inspectors do not look for documentation of this type of training because they rely on interviews with employees about their job-specific safety knowledge, for such a wide range of tasks.

C. UPDATE

There is no requirement for testing or quantifying staff competencies before and/or after training. However, as much as 13% of the employee population can be interviewed within the scope of a normal inspection. During those interviews, inspectors will ask employees about their safety training and their knowledge. (See Section III. This is a list of the questions used in the employee interview. The extent and nature of the interviews will depend on the findings on the facility's OSHA 200 Log, the document review process, and the walk-through.)

D. CONTENTS

The "OSHA Mandatory Training Compliance Cross-Reference Tracker" (Section II) is a guide designed to maximize the OSHA credit and benefit a facility can reap from the training conducted. It bridges the inconsistency that often exists between the OSHA method of document review as mentioned above and typical facility records, which are in chronological order and by individual personnel records. The guide provides a discussion on the specific training requirements of the basic eight, or nine to ten (depending on your operations), OSHA mandatory subjects. It also provides a cross-reference of frequently used long-term care titles to the OSHA standards to facilitate alignment of in-house training titles with OSHA terminology. This improves the likelihood for inspector recognition and, thus, facility credit for training conducted. This tool will also help to identify training deficiencies quickly.

The OSHA inspection employee interview questions in Section III is a copy of these guides and Excel chart sheets, entitled "EmployeeInterview.doc," "Training.doc," and "Training.xls" and are available at the CRC Web site.

II. OSHA MANDATORY TRAINING COMPLIANCE CROSS-REFERENCE TRACKER

A. INTRODUCTION

This tool was developed to help long-term care facilities show compliance with required OSHA training subjects. The employee training records of most long-term care facilities are organized chronologically, and by individual employee reference. However, OSHA seeks training information *by subject.*

The following pages can be used as a facility's "Training Cross-Reference Tracker." This guides the user to reap as much credit as possible for the training that has been conducted throughout the year. It helps to identify and highlight mandatory issues that may have been covered under, or along with, other subject titles, which are not as obvious or relevant to OSHA compliance. It will most likely be kept in the beginning of the training documentation binder for easy reference and recording.

B.   OSHA TRAINING REQUIREMENT

Remember, OSHA is only concerned with training to protect *employee* health and safety. Resident rights and safety are only marginally relevant to the extent that they affect employee safety and health.

To show compliance with OSHA requirements, one must be able to produce the following:

Evidence of *annual training* of required subjects—schedule and signed rosters
Evidence that *appropriate employees received appropriate training*—signed rosters with title and department, random review of employee training records exhibiting title and department along with corresponding training received
Evidence of *Content and Materials Covered and Distributed in Training*

[Not required but strongly recommended purely as an operational measure, and not a compliance measure, at least not yet.]
Training markers—quick quizzes before and after in-service to evaluate need, impact, and effectiveness of training given

C.   OSHA REQUIRED TRAINING SUBJECTS

- Bloodborne pathogens
- Tuberculosis
- Fire and emergency
- Fire extinguisher
- Lockout/tagout
- Personal protective equipment
- Hazard communication (eyewash)
- Ergonomics
- (Respiratory protection)
- Workplace violence prevention (behavior management)

D.   TRANSLATING OSHA TRAINING REQUIREMENTS TO A FACILITY'S EXISTING TRAINING TITLES

Most long-term care facilities generally provide adequate training to fulfill OSHA requirements. Differences in semantics and terminology make showing compliance a little cumbersome at times. This tool will help clarify terms, material content, and expectations to minimize the amount of confusion and misunderstanding for a facility and the OSHA inspector. It will help organize training information optimally for an OSHA inspection. This tool was designed under the assumption that training records are kept chronologically, by date order, on an annual basis, with each entry to include a basic outline, sign-in sheet roster, and sample training content and materials in very close proximity (same binder or file, albeit not necessarily repeatedly entered for each entry).

Listed below are the OSHA Mandatory Training Subjects along with a list of other potential subjects, titles, and topics, which are equivalent, or included in such titles. Review training records, and use the attached spreadsheet to note dates of training that was conducted, which could fulfill mandatory training requirements, and indicate how much of the training given was dedicated to the subject (e.g., full, half, part, or actual time in minutes).

How a facility uses this tool is entirely up to it based on the requirements delineated in the preceding pages. Adequate coverage of certain topics is dependent on the gravity and the subject itself. For example, it would not suffice to have 15 minutes on hazard communication, TB, or bloodborne pathogens, whereas that may be feasible for reviewing eye washes, or personal protective equipment as part of a larger topic on hazard communication or bloodborne pathogens. The whole idea is to garner as much credit as possible for the training that is done, and to organize it in such a way that the information is readily accessible.

General Health and Safety

Although there is no specific requirement for general safety training, it is a good place to start. It shows "good faith" efforts, and organizes and highlights a facility's miscellaneous safety training. Any health and safety items that cannot be fit anywhere else should be highlighted in this miscellaneous section (e.g., general safety training during orientation, machine guarding, hand-tool safety, etc.). Here is an opportunity to take as much credit as possible for any and all safety training and activities.

Hazard Communication (also known as Hazcom)

Worker right-to-know (Hazcom is also known as Worker Right-to-Know Law)
Chemical safety
MSDS training

Ergonomics

Proper lifting techniques
Patient transfer
Patient handling
Back care
Back safety
Lift equipment training (Hoyer, Arjo, Versa, etc.)
Remedial lifting procedures training for injured employees

Bloodborne Pathogens

Standard precautions (old universal precautions)
Handling medical waste
Bloody linens
Handling sharps/sharps containers
Handwashing
Use of gloves and aprons
Needle sticks/needle stick prevention
Isolation
HIV/HBV
HBV vaccine
Infection control

Tuberculosis

Infection control
Exposure control
Respiratory protection
Negative air isolation
PPD

Workplace Violence Prevention

Behavior management
Dealing with combative residents
Coping with behavior problems
Dementia/Alzheimer's disease
Special care programs/units
In-Services or discussions on providing information on shift logs, quality assurance, and care planning meetings on resident psychosocial status, level of irritability, potential for acting out, history of combativeness, etc.; any behavior management activities communicated to employees for their own edification
Robbery/civil disobedience/hostage situations, etc.

Fire and Emergency

Fire plan
RACE (Remove, Alarm, Contain, Evacuate)
Disaster plan

Fire Extinguisher

Types of fire extinguishers (and their use)
How and when to use a fire extinguisher

Eyewash

Eyewash stations/stands/bottles locations, how and when to use them

Note: Employees are expected to be trained on all equipment, safe job tasks, and safety equipment but documentation is not usually sought. However, since eyewashes have been a routine issue on long-term care inspections, it behooves the facility to document that training has been provided for this specific piece of safety equipment.

Lockout/Tagout

Control of energized equipment
Electrical safety (must include lockout/tagout procedures)
The lockout/tagout policy

Note: Affected employees (staff working around or near lockout/tagout work being done), and not only authorized employees (staff carrying out the work that requires to be locked and tagged out), require annual lockout/tagout training. Therefore, all staff should receive annual lockout/tagout training. Training for affected employees is very basic. It only needs to cover the highlights of the program to make the employees aware of its existence, and what their co-workers, the authorized employees, are actually doing.

Personal Protective Equipment (PPE)

Gloves (latex, leather work gloves, rubber, etc.)
Goggles
Aprons
Face shields
Respirators
Hard hats
Not back belts (back belts are not considered PPE)

Respiratory Protection

> Only necessary if respirators used for either TB or chemicals
> Fit testing
> Respiratory protection program
> Respirator training

See Appendix C for sample charts.

III. OSHA INSPECTION EMPLOYEE INTERVIEW QUESTIONS

OSHA inspections usually begin with a document review and then proceed to a site inspection, which also includes employee interviews. Inspectors can interview a maximum of 13% of the employee population. If the facility is unionized, then the inspector may ask the union representative available to select the personnel to be interviewed. Otherwise, the inspector will select staff randomly. There would be a mix of nursing, housekeeping, dietary, and maintenance personnel. Sometimes the physical therapist may also be "put on the spot" because of their involvement in ergonomics. Interviews are generally not extensive unless issues are discovered or suspected to exist. No management personnel is allowed to be present during the interviews. Interviews are conducted in private.

Following are some sample questions that may be asked. A quick glance may start your imagination on how questions, issues, and curiosity can develop on the inspection. Review these questions to ensure that responses will coincide with appropriate policies and training provided. These questions can also be used as an in-house survey to verify training needs or to identify safety management system issues that need to be addressed. Documentation of well-implemented programs, along with good training records, is also essential to defend the facility should employee responses be inadequate.

In the inspection preparedness section, the Inspection Policy Guide suggests that the facility liaison with OSHA speak with interviewed employees immediately after they have been interviewed. The liaison should verify what was asked, and how they answered, and document it on the running inspection log. This is an excellent way to clear up any misunderstandings and nip issues in the bud before they bloom out of control. It helps ensure that the employee(s) understood the inspector's questions; ensure that you understood the employee's responses; verify if the inspector did as well; and begin to address issues immediately that have legitimately surfaced to stave off further misunderstanding or findings of deficiencies.

This list of questions, entitled "EmployeeInterview.doc," is available in survey format on the CRC Web site; a sample is provided on the next pages.

SAMPLE: SURVEY QUESTIONS

Have you had, or do you know of anyone who has had, a workplace accident? Tell me about it. (Information will be verified with OSHA 200 Logs and accident reports.)

_____

_____

_____

Did you ever, or do you know of anyone who has ever, sustained a bloodborne "exposure incident"? For example, a needle stick or bodily fluids with visible blood splashed to the face/mucous membrane?

_____

_____

_____

Do you know what an MSDS is?

_____

_____

_____

Do you know where the MSDSs are stored in your department?

_____

_____

_____

What are some products you use? What kind of personal protective equipment do you use with them? What do you do in case of a spill or chemical exposure?

_____

_____

_____

Do you know where the eyewashes are in your work area? Do you know how to use them? Please show me.

_____

_____

_____

Did you, and how often do you, get training on Hazard Communication? TB? Bloodborne Pathogens? Body Mechanics? Emergency and Disaster Plan? Fire Extinguisher Use? Workplace Violence Prevention?

_____

_____

_____

What do you do in case of a fire? Where do you evacuate to?

_____

_____

_____

Has there ever been anyone here who was suspected of having TB? What happened? How was it handled?

_____

_____

_____

Is there a Health and Safety Committee? Have you ever participated in anything it has done? What does it do? Is it effective? How does it help you?

_____

_____

_____

Is there a Health and Safety Program? Is it effective?

_____

_____

_____

What happens if you find a serious safety hazard? How do you report it? What would happen? How are employee safety complaints addressed here?

_____

_____

_____

How do you handle infectious waste? Soiled linens? Infectious soiled linens?

_____

_____

_____

Are there enough sharps containers? Are they in accessible places? Do they ever get filled over the top? How are they emptied or disposed of?

_____

_____

_____

Do you ever lift alone? When?

_____

_____

_____

Do you feel there is adequate lift equipment and training in patient transfer techniques?

_____

_____

_____

Have you, or do you know of any staff who have, ever been assaulted by a resident? What happened? How was it handled?

_____

_____

_____

Are you provided with adequate and proper personal protective equipment?

_____

_____

_____

Do you practice lockout/tagout procedures? When? How? Show me.

_____

_____

_____

Appendix A

Risk Map of a Long-term Care Facility with Potential Hazards

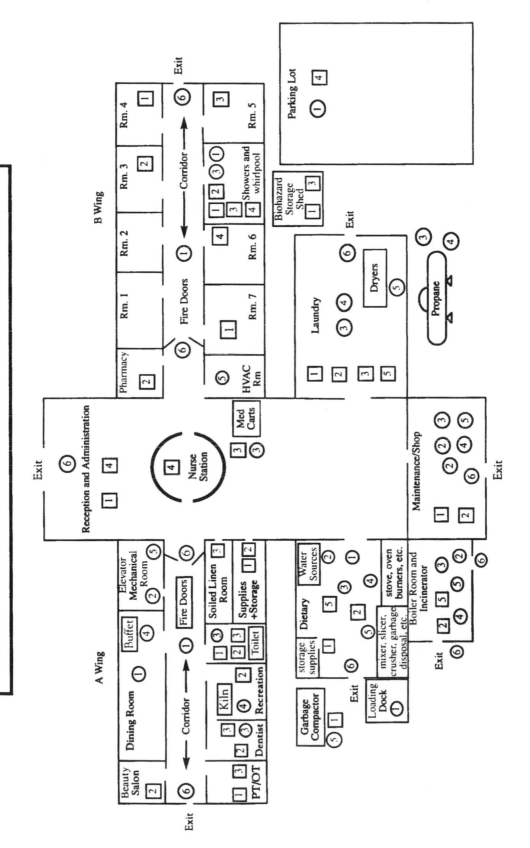

RISK MAP OF A LONG-TERM CARE FACILITY WITH POTENTIAL HAZARDS

Some Potential Safety Hazards

1. Slips and Falls (walking working surfaces, heights)
2. Electrical Shock (electrical safety, lockout/tagout)
3. Personal Protective Equipment
4. Burns/Fire/Explosion
5. Machine Guarding (lockout/tagout, guards)
6. Egress (obstruction)

Some Potential Health Hazards

1. Lifting and Ergonomic Hazards (lifts, training, dollies, etc.)
2. Chemical Exposure (PPE, MSDS, Hazcom, Eyewash)
3. HBV, Bloodborne, Sharps, TB (PPE, ECP)
4. Workplace Violence (training+)
5. Heat Stress

Appendix B

Job Safety and Health Protection

JOB SAFETY & HEALTH PROTECTION

The Occupational Safety and Health Act of 1970 provides job safety and health protection for workers by promoting safe and healthful working conditions throughout the Nation. Provisions of the Act include the following:

Employers

All employers must furnish to employees employment and a place of employment free from recognized hazards that are causing or are likely to cause death or serious harm to employees. Employers must comply with occupational safety and health standards issued under the Act.

Employees

Employee must comply with all occupational safety and health standards, rules, regulations and orders issued under the Act that apply to their own actions and conduct on the job.

The Occupational Safety & Health Administration (OSHA) of the U.S. Department of Labor has the primary responsibility for administering the Act. OSHA issues occupational safety and health standards, and its Compliance Safety and Health Officers conduct jobsite inspections to help ensure compliance with the Act.

Inspection

The Act requires that a representative of the employer and a representative authorized by the employees be given an opportunity to accompany the OSHA inspector for the purpose of aiding the inspection.

Where there is no authorized employee representative, the OSHA Compliance Officer must consult with a reasonable number of employees concerning safety and health conditions in the workplace.

Complaints

Employees or their representatives have the right to file a complaint with the nearest OSHA office requesting an inspection if they believe unsafe or unhealthful conditions exist in their workplace. OSHA will withhold, on request, names of employees complaining.

The Act provides that employees may not be discharged or discriminated against in any way for filing safety and health complaints for otherwise exercising their rights under the Act.

Employees who believe they have been discriminated against may file a complaint with their nearest OSHA office within 30 days of the alleged discriminatory action.

Proposed Penalty

The Act provides for mandatory civil penalties against employers of up to $7,000 for each serious violation and for optional penalties of up to $7,000 for each nonserious violation. Penalties of up to $7,000 per day may be proposed for failure to correct violations within the proposed time period and for each day the violation continues beyond the prescribed abatement date. Also, any employer who willfully or repeatedly violates the Act may be assessed penalties of up to $70,000 for each such violation. A minimum penalty of $5,000 may be imposed for each willful violation. A violation of posting requirements can bring a penalty of up to $7,000.

There are also provisions for criminal penalties. Any willful violation resulting in the death of any employee, upon conviction, is punishable by a fine of up to $250,000 (or $500,000 if the employer is a corporation), or by imprisonment for up to six months, or both. A second conviction of an employer doubles the possible term of imprisonment. Falsifying records, reports, or applications is punishable by a fine of $10,000 or up to six months in jail or both.

Voluntary Activity

While providing penalties for violations, the Act also encourages efforts by labor and management, before an OSHA inspection, to reduce workplace hazards voluntarily and to develop and improve safety and health programs in all workplaces and industries. OSHA's Voluntary Protection Programs recognize outstanding efforts of this nature.

OSHA has published Safety and Health Program Management Guidelines to assist employers in establishing or perfecting programs to prevent or control employee exposure to workplace hazards. There are many public and private organizations that can provide information and assistance in this effort, if requested. Also, your local OSHA office can provide considerable help and advice on solving safety and health problems or can refer you to other sources for help such as training.

Consultation

Free assistance in identifying and correcting hazards and in improving safety and health management is available to employers, without citation or penalty, through OSHA supported programs in each State. These programs are usually administered by the Sate Labor or Health department or a State university.

Citations

If upon inspection OSHA believes an employer has violated the Act, a citation alleging such violations will be issued to the employer. Each citation will specify a time period within which the alleged violation must be corrected.

The OSHA citation must be prominently displayed at or near the place of alleged violation for three days, or until it is corrected, whichever is later, to warn employees of dangers that may exist there.

Posting Instructions

Employers in States operating OSHA approved State Plans should obtain and post the State's equivalent poster.

Under provisions of Title 29, Code of Federal Regulations, Part 1903.2(a)(1) employers must post this notice (or facsimile) in a conspicuous place where notices to employees are customarily posted

More Information

Additional information and copies of the Act, OSHA safety and health standards, and other applicable regulations may be obtained from your employer or from the nearest OSHA Regional Office in the following Locations:

Atlanta, GA	(404) 347-3573
Boston, MA	(617) 565-9860
Chicago, IL	(312) 353-2220
Dallas, TX	(214) 767-4731
Denver, CO	(303) 844-1600
Kansas City, MO	(816) 422-5861
New York, NY	(212) 337-2378
Philadelphia, PA	(215) 596-1201
San Francisco, CA	(415) 975-4310
Seattle, WA	(206) 553-5930

Washington, DC
1996 (Reprinted)
OSHA 2203

Robert B. Reich, Secretary of Labor

U.S. Department of Labor

Occupational Safety and Health Administration

This information will be made available to sensory impaired individuals upon request. Voice phone: (202) 219-8615; TDD message referral phone: 1-800-326-2577.

Appendix C

OSHA Mandatory Compliance Cross Reference Tracker

OSHA MANDATORY COMPLIANCE CROSS REFERENCE TRACKER

Year: _____   Facility: _____   Sheet No: _____

OSHA Training Subjects:	Entry#	Date Of In-Service	Duration (Check one, or)		(ex:10min/ 30 min)	Department(s)	Alternative Title/Specific Items Covered:		
			Full	Half	Part	min/min				
Fire Extinguisher										

OSHA MANDATORY COMPLIANCE CROSS REFERENCE TRACKER

Year: _____ Facility: _____ Sheet No: _____

OSHA Training Subjects:	Entry#	Date Of In-Service	Duration (Check one, or) Full	Half	Part	(ex:10min/ 30 min) min/min	Department(s)	Alternative Title/Specific Items Covered:
Fire Extinguisher								

OSHA MANDATORY COMPLIANCE CROSS REFERENCE TRACKER

Year: _____ Facility: _____ Sheet No: _____

OSHA Training Subjects:	Entry#	Date Of In-Service	Duration (Check one, or) Full	Half	Part	(ex:10min/ 30 min) min/min	Department(s)	Alternative Title/Specific Items Covered:
Respiratory Protection Program (only if using respirators, see also Tuberculosis and PPE section)								

OSHA MANDATORY COMPLIANCE CROSS REFERENCE TRACKER

Year: _____   Facility: _____   Sheet No: _____

OSHA Training Subjects:	Entry#	Date Of In-Service	Duration (Check one, or)			(ex:10min/ 30 min) min/min	Department(s)	Alternative Title/Specific Items Covered:
			Full	Half	Part			
Eye Wash and Eye Protection								

OSHA MANDATORY COMPLIANCE CROSS REFERENCE TRACKER

Year: _____ Facility: _____ Sheet No: _____

OSHA Training Subjects:	Entry#	Date Of In-Service	Duration (Check one, or) Full	Half	Part	(ex:10min/ 30 min) min/min	Department(s)	Alternative Title/Specific Items Covered:
Fire Extinguisher								

Appendix B

Job Safety and Health Protection

OSHA MANDATORY COMPLIANCE CROSS REFERENCE TRACKER

Year: _____   Facility: _____   Sheet No: _____

OSHA Training Subjects:	Entry#	Date Of In-Service	Duration (Check one, or) Full	Half	Part	(ex:10min/ 30 min) min/min	Department(s)	Alternative Title/Specific Items Covered:
Lockout/Tagout								

OSHA MANDATORY COMPLIANCE CROSS REFERENCE TRACKER

Year: _____ Facility: _____ Sheet No: _____

OSHA Training Subjects:	Entry#	Date Of In-Service	Duration (Check one, or) Full	Half	Part	(ex:10min/ 30 min) min/min	Department(s)	Alternative Title/Specific Items Covered:
Fire Plan								

OSHA MANDATORY COMPLIANCE CROSS REFERENCE TRACKER

Year: _____   Facility: _____   Sheet No: _____

OSHA Training Subjects:	Entry#	Date Of In-Service	Duration (Check one, or) Full	Half	Part	(ex:10min/ 30 min) min/min	Department(s)	Alternative Title/Specific Items Covered:
Hazard Communication								

OSHA MANDATORY COMPLIANCE CROSS REFERENCE TRACKER

Year: _____   Facility: _____   Sheet No: _____

OSHA Training Subjects:	Entry#	Date Of In-Service	Duration (Check one, or) Full	Half	Part	(ex:10min/ 30 min) min/min	Department(s)	Alternative Title/Specific items Covered:
Bloodborne Pathogens								

OSHA MANDATORY COMPLIANCE CROSS REFERENCE TRACKER

Appendix C: Training Record Sheets

Year: _____ Facility: _____ Sheet No: _____

OSHA Training Subjects:	Entry#	Date Of In-Service	Duration (Check one, or)		(ex:10min/ 30 min)	Department(s)	Alternative Title/Specific Items Covered:
			Full	Half	Part	min/min	
General Health and Safety								

Appendix D

OSHA Inspection Preparedness Checklist for Long-Term Care Facilities

These are in basic chronological order. Each element is critical.

I. OSHA INSPECTOR'S INTRODUCTION

1. Is the receptionist aware of how to handle an OSHA inspector?
2. Do you verify the inspector's identification?
3. Do you have the inspector sign in?
4. Does the receptionist know who to relay the inspector to? And who else to relay the inspector to if the primary contact is not available?
5. Is the facility contact aware of how to handle an OSHA inspector?
6. Do you admit entry without a warrant?
7. Do you offer coffee and doughnuts?
8. Do you verify the purpose and scope of the inspection?
9. Do you provide the inspector with a comfortable and quiet space to work in for the duration of the inspection?
10. Do you approach and deal with the inspector with respect and professional courtesy?
11. Do you only answer and provide information that is requested?
12. Do you refrain from offering more information than is requested?
13. Do you attempt to present your case or attempt to clarify that you believe that you may actually be in compliance, if the inspector suspects or states that you are not, when the issue warrants?
14. Do you show the inspector corrected item(s) that were cited before the inspector leaves the site?

II. OSHA 200 LOG—AVAILABLE, COMPLETE, AND ACCURATE

1. Are the OSHA 200 Logs readily available for the years 1993 to present?
2. Are the logs kept in one central location and are they well organized?
3. Do the logs appear to be correctly recorded, with appropriate check marks in the columns on the right, and tallies on the bottom?
4. Are incidents that are known to be recordable recorded within 6 days of knowledge?
5. Are cases properly recorded?
6. Are all loss-time cases recorded?
7. Does the log only contain injuries sustained by employees, not volunteers, residents, visitors, or vendors?
8. Are all cases truly work related (not parking lot slips, lunch outing car accidents, etc.)?
9. Are all "first aid" cases based on the OSHA definition of first aid (no time loss, negative radiograph, initial application of cold, heat, or whirlpool treatment, initial dose of prescription drug, treatment of minor cuts, bruises, scratches, and burns, etc.) determined to be nonrecordable and, thus, do not appear on the log?
10. Are all "medical treatment" cases based on the OSHA definition of medical treatment (loss time, loss of consciousness, stitches, more than one application or dose of prescription medication, heat, cold, or whirlpool treatment, third degree burns, etc.) deemed recordable and, thus, appear on the log, even if there is no loss time?
11. Are TB conversions from a baseline negative recorded on the log on the "illnesses" portion?
12. Are all potential bloodborne exposures (i.e., needle stick) recorded on the log?
13. Are there referring accident reports for all cases listed?
14. Are there referring workers' compensation records for those cases?

15. Is the log properly tallied, signed, and is the bottom right-hand corner of the log posted in an employee area for the month of February?
16. Is the information from the log shared with safety committee and those with safety management responsibilities?
17. Does the administrator review the log for accuracy on a periodic basis?

III. HEALTH AND SAFETY PROGRAM

1. Does a written health and safety policy exist?
2. Does it describe the safety system and the mission goals?
3. Does it define the safety committee, its activities, chain of accountability?
4. Does it serve as an umbrella policy via reference to all other health and safety policies delineating personnel, duties, and activities for each subject (i.e., orientation safety and periodic safety training, hazard communication, bloodborne pathogens, TB, ergonomics, workplace violence, prevention, etc.)?
5. Does an employee health and safety committee exist?
6. Are employees aware that a health and safety program exists?
7. Do employees feel that the health and safety program is responsive?
8. Does at least half of the safety committee consist of employee representation?
9. Are employee participants on the safety committee tenured for at least a year?
10. Are employees provided training so that their participation in these meetings may be more meaningful and productive?
11. Is there any log or other means of tracking safety activities, endeavors, issues, and accomplishments (outside of meeting minutes) to reflect activities and achievements?
12. How are safety activities communicated to employees?
13. Are safety committee minutes posted in an employee area?
14. Is top management directly involved in the safety committee and safety program?
15. Is top management visible and accessible to employees on safety issues and activities?
16. Is top management a role model to employees on safety issues and activities?
17. Is top management committed to the goals and mission of both?
18. Is the program reviewed annually and updated and signed?

IV. HAZARD COMMUNICATION

1. Does a written Hazcom policy exist?
2. Does it address the management of chemical safety throughout the facility?
3. Does it refer to a master chemical inventory list of all chemicals on site in alphabetic order by product name?
4. Does it refer to a master MSDS (Material Safety Data Sheet) file of all substances that are used on site, and have been used on site in the past?
5. Does it refer to subfiles of MSDSs that contain only current MSDSs relevant to the area or department, where they are readily available?
6. Does it refer to a uniform chemical labeling system?
7. Does it refer to appropriate annual employee training?
8. Does it have a provision for covering contractors?
9. Are all containers properly labeled throughout the facilities?
10. Do employees use appropriate personal protective equipment when working with chemicals?
11. Do employees know emergency and cleanup procedures?

12. Do employees know what an MSDS is, and where they are kept?
13. Are chemicals stored and used in such a way as to reduce the risk of leakage, spillage, interactivity, and exposure?
14. Is the policy reviewed periodically, signed, and updated?

V. BLOODBORNE PATHOGENS—EXPOSURE CONTROL PLAN

1. Does your exposure control plan assess risks by job categories?
2. Does it provide for standard precautions procedures?
3. Does it address personal protective equipment needs?
4. Does it establish the use of the biohazard signage?
5. Does it provide guidance for assessing and managing exposures including accident investigation, medical follow-up, counseling, including new considerations brought upon by recent CDC 6/7/97 guideline revisions on HIV exposure treatment?
6. Is postexposure follow-up medical documentation, including medical counseling records, source patient identification, blood archiving, etc., available on site?
7. Does it address the provision for tracking hepatitis B vaccinations?
8. Does the facility use the mandatory terminology for employee declination of vaccine?
9. Does the plan address annual employee training?
10. Does it clarify what constitutes an "exposure incident," and what does not?
11. Does it address reportability of exposure incidents on the OSHA 200 Log and the lines of communication necessary to ensure that exposure incidents are properly recorded and/or followed?
12. Are sharps containers properly used and maintained so that they are never overfilled?
13. Are sharps containers, medical wastes, and red bag laundry stored in *secured* and properly labeled rooms (with the proper biohazard label)?
14. Do employees practice standard precautions?
15. Does the administration follow the postexposure follow-up procedures whenever there is an exposure incident?
16. Is the plan reviewed periodically, updated, and signed?

VI. TUBERCULOSIS—INFECTION CONTROL PLAN

1. Does the infection control plan assess risk according to status and profiles of residents, staff, and the community?
2. Does it provide for annual employee training?
3. Does it provide for protocols for identifying, isolating, and discharging residents with suspect or active TB?
4. Does it provide for a respiratory protection program to include the use of appropriate masks, staff selection, fit testing, and training, if respirators are on site?
5. Is there an accompanying written respiratory protection program if respirators are present?
6. Does the plan provide for temporary negative pressure isolation room setup and procedures to be used in conjunction with the respiratory protection program? (necessary unless a facility is strict about following discharge procedures for "suspect" residents, and sure that there will be no delay in the discharge by receiving hospital.)
7. Does it provide for medical removal for employees with suspect or active TB?
8. Are employees aware of TB symptoms, and that they should report them immediately if they experience them, or believe a resident has those symptoms?
9. Do employees receive PPD tests annually?

10. Do residents receive PPD tests every other year?
11. Does the plan provide for recording any baseline employee conversions from a negative to a positive on the illness side of the OSHA 200 Log?
12. Is the plan reviewed periodically, updated, and signed?

VII. ERGONOMIC PROGRAM

1. Is there an ergonomic policy?
2. Does it monitor, evaluate, and address ergonomics for *all* employees, not just nursing?
3. Does it address nursing lift equipment needs? Is there an adequate number of these?
4. Does it address ergonomic training needs of all departments?
5. Does it provide for a system of determining optimal ergonomics in purchasing of equipment, furniture, devices, and in designing area plans, *workstations,* and *work flow*?
6. Is there adequate training on the use of these devices?
7. Are assistive lifting devices and other equipment properly maintained?
8. Do employees practice good body mechanics? Do they follow the care plan?
9. Are employees coached to ask for assistance, and to provide assistance whenever necessary?
10. Are employees properly disciplined for deviating from established lifting procedures whenever it occurs, regardless whether an injury is precipitated by the event?
11. Are laundry carts and meal carts designed for better ergonomics? (lighter construction, larger and softer constructed wheels, design allows for optimal visibility, laundry carts may have spring action to reduce bending as load decreases, etc.)
12. Is storage designed for optimal ergonomics? (heavier, bulkier, and larger items stored at waist level or lower, and lighter smaller items stored at or above shoulder height; also, storage at waist height as much as possible).
13. Is the program reviewed periodically, updated, and signed?

VIII. LOCKOUT/TAGOUT PROGRAM

1. Does a written lockout/tagout policy exist?
2. Does it have a provision for contractors?
3. Does it provide for the *annual in-house inspection and certification* by the administrator that the program is in order and that all necessary corrective measures have been taken?
4. Does it provide for annual training of authorized *and affected* employees?
5. Does it provide for a comprehensive evaluation of all equipment and operations that may require lockout for all or certain repairs or maintenance work?
6. Does it provide protocols for appropriate lockout procedures?
7. Is there appropriate and adequate lockout equipment?
8. Is it routinely used?
9. Are other staff aware of the lockout/tagout program?

IX. EMERGENCY AND DISASTER PLAN

1. Does it cover all employees?
2. Does it instruct employees on where they should go in the event of an evacuation?
3. Is there a system of accounting for all employees in and out of the building?
4. Is there annual fire extinguisher training (if there is no formal policy stating that no employee is to use a fire extinguisher)?

5. If there is a formal policy against employees using fire extinguishers, is there annual training on this policy? Do employees really know not to use one, in this case?
6. Is the plan reviewed periodically, updated, and signed?

X. WORKPLACE VIOLENCE PREVENTION POLICY

1. Does a comprehensive workplace violence prevention policy exist?
2. Does it assess community, staff, and resident risk, based on available local resources, past experience, layout, logistics, lighting, existing security measures, etc.?
3. Does it define acceptable and unacceptable conduct in terms of physical and nonphysical violence? (May make reference to tie into sexual harassment, horseplay, and other policies on inappropriate behaviors and conduct.)
4. Does it address the possession of weapons or firearms?
5. Does it address employees with restraining orders on spouses or stalkers?
6. Does it provide for annual employee training on this policy?
7. Does the training include identifying potentially violent situations and how to manage them?
8. Is the policy reviewed periodically, updated, and signed?

XI. CONFINED SPACES ASSESSMENT

1. Has there been a survey of the facility to identify all potential "confined spaces" (usually elevator pits and cesspool pits for septic systems)?
2. Are there proper signs labeled "Confined Space—Restricted Entry" on doors or entryways to confined spaces?
3. Are these doors or entryways secured?
4. Is it assured that the elevator contractor practices appropriate lockout procedures to render the elevator pit as not a "permit-required confined space" by removal of the mechanical hazard?
5. Is it assured that contractors who must enter septic systems have appropriate safety equipment and air monitoring equipment?
6. Is the assessment reviewed, updated, and signed periodically?

XII. EMPLOYEE TRAINING RECORDS

1. Do records reflect annual mandatory training on OSHA subjects and adequate attendance?
2. Are they clear and readily available for review?
3. Are they in a central location?
4. Do subjects include bloodborne pathogens, TB, hazard communication, emergency plan, fire extinguisher use, ergonomics, workplace violence, including dealing with dementia/behavior management, and lockout/tagout?
5. Do records reflect class plan, content, materials covered and qualification of instructor?

XIII. GENERAL ELECTRICAL SAFETY

1. Are GFI (ground fault interrupters) installed near (6 ft) water sources (kitchens/bathrooms)?

2. Is there absolutely no use of extension cords except for temporary use, i.e., holiday lights (resident room use for fans and such is not acceptable)?
3. Are all ground plugs in working order on appliances?
4. Are all appliances checked for missing ground plug, frayed wires, and overall electrical integrity of the appliances?
5. Are all electrical boxes properly covered and closed?
6. Is there any exposed wiring, boxes, electrical or utility equipment (telephone and computer lines are of such low voltage they are not covered)?
7. Are all heavy machinery (boilers, etc.) properly hardwired?
8. Are all electrical installations designed to support the amperage supplied?

XIV. MACHINE GUARDING

1. Are all moving parts under 7 ft tall guarded?
2. Are the backs of laundry dryers properly guarded?
3. Are slicers properly guarded?
4. Are Hobart mixers (and other machines with moving parts) properly guarded?
5. Are all fans properly guarded (no more than one-third in. distance to guard)?
6. Are bench grinders, punch presses, saws, etc. properly guarded with available eye and hand protection?
7. Emergency generators may not be properly guarded as delivered by the manufacturer, particularly older models. Is such equipment also surveyed for proper machine guarding?
8. Are screens on blowers, fans, compressors intact?
9. Is all broken equipment (including broken screens and guards) locked out and tagged with a sign "out of service"?
10. Are garbage disposals, garbage compactors, can crushers, meat slicers, etc. properly guarded, operated, and maintained to prevent amputation?

XV. PREVENTATIVE MAINTENANCE

1. Is there a comprehensive preventative maintenance program for the HVAC, safety, security, and life support systems?
2. Is there a comprehensive preventive maintenance and monitoring program for equipment, lift devices, electrical appliances, eyewashes (monthly), etc.?
3. Is this program well documented?
4. Is there a good communication system wherein any employee can report a maintenance problem that will be addressed appropriately in a timely fashion?

XVI. EYEWASH

1. Are ANSI (American National Standards Institute)-approved, plumbed (connected to water source by piping) eyewashes available in areas of reasonable risk of chemical or other exposures to the eye?
2. Is the fixture marked with a green safety eyewash sign?
3. Is the fixture within 100 ft or 10 seconds running distance of the exposure area?
4. Are the eyewashes flushed and checked monthly?
5. Is the monthly check logged?
6. Are employees given training on the locations, purpose, and operation of the eyewashes?

XVII. WALKING AND WORKING SURFACES AND MISCELLANEOUS

1. Are walking and working surfaces properly designed, installed, and maintained to prevent slips, trips, and falls?
2. Are rubber mats available in heavily trafficked wet areas, i.e., kitchens? Or is some other alternative friction enhancement system available?
3. Are drop-off points well designated with markers and/or chain rope to prevent falls (i.e., loading platform, stairs)?
4. Are all substantial bumps, holes, or surface irregularities marked and managed in such a way to minimize the potential for injury (i.e., drain holes, skylights, sudden grade changes, etc.)?
5. Do all working areas have adequate lighting?

XVIII. SAFETY INCENTIVE

1. Are safety incentive programs, systems, or any type of awareness or motivational program *performance driven* (based on desired behaviors), and not "number of incident" driven?
2. Do supervisors and managers provide proper motivation, training, coaching, along with and in greater proportion to discipline?
3. Are safety infractions of any kind appropriately and consistently disciplined?

XIX. MISCELLANEOUS

1. Are all means of egress free of obstruction?
2. Are employees using proper personal protective equipment?
3. Are employees practicing safe work procedures?
4. Are all compressed gas cylinders proper stored and used?
5. Is the facility asbestos-free?

Appendix E

Files for Downloading on CRC Web Site

The vast majority of the sample policies, guides, checklists, and questionnaires presented in this book are available online. They are mostly in Word, and some are in Excel. Files ending with ".doc" are Word files and files ending with ".xls" are Excel files. The following is an index of these files available at the CRC Web site at: http://www.crcpress.com.

Document Name	File Name
OSHA Inspection Policy Guide	InspectionPolicy.doc
The body of the OSHA Program Evaluation Profile	PEP.doc
Sample Health and Safety Policy	Health&SafetyPolicy.doc
Health and Safety Activities (Chart)	Health&SafetyActivities Log.doc
	Health&SafetyActivities Log.xls
Asbestos—Reinspection/Job Request/Authorization Forms	AsbestosGuide.doc
Access to Exposure and Medical Records Sample Letter	AccessMed&ExpRec.doc
Bloodborne Guide	BloodborneGuide.doc
Confined Space Guide	ConfinedSpace.doc
Hazard Communication Guide	HazcomGuide.doc
Outline	(same file)
Sample Hazard Communication Policy	(same file)
Sample MSDS Inquiry Letter	(same file)
MSDS Glossary	(same file)
Lockout/Tagout Sample Policy Guide	LockoutTagoutGuide.doc
Personal Protective Equipment	PPE.doc
Sample Ergonomic Policy Guide	ErgonomicGuide.doc
Tuberculosis Management Guide	TBGuide.doc
Practical Guide	(same file)
CDC/OSHA Synthesis Summary	(same file)
CDC Definitions of Five Levels of Risk	(same file)
TB Program Checklist	(same file)
TB Respiratory Protection Program Sample Policy	TBRPPPolicy.doc
Sample Workplace Violence Prevention Policy Guide	WorkplaceViolence.doc
Sample Resident Behavior Management Plan	BehaviorMgmt.doc
Mandatory Training Cross-Reference Tracker Guide	Training.doc
(Charts)	Training.xls
Employee Interview Questions	EmployeeInterview.doc
Comprehensive Inspection Preparedness Checklist	Checklist.doc
OSHA 200 Log of Injuries and Illnesses Guide	OSHA200Log.doc
List of Agencies	ListofAgencies.doc

Appendix F

OSHA 200 Log of Injuries and Illnesses Guide

A. GOAL

OSHA relies on statistics gathered from these logs to identify high hazard industries, job categories, etc., to plan and target special emphasis programs and to inspect a certain percentage of an industry that has high average injury rates or disturbing outcomes. OSHA inspectors begin the inspection by reviewing the OSHA 200 Log to verify its accuracy and to determine the level and type of risks and accident history at the site.

B. GENERAL SUMMARY

- The official form is free and is available from local OSHA offices. Users are allowed to make copies of an original form. Users are also allowed to use a different form, as long as it is as *"detailed, easily readable and understandable as the OSHA 200 Log."* (This includes some OSHA 200 Log recordkeeping software print outs available on the market from private companies.) Users can also alter the original size of the OSHA 200 Log form. Commonly, legal size is used for more convenient duplication, as long as it is kept as *"detailed, easily readable and understandable."*
- Logs must be maintained for 5 years.
- There should be a separate Log for each calendar year.
- The information on the Log should be supported and should match the information derived from in-house accident report forms, workers' compensation records, employee medical records, and responses from employee interviews.
- Accuracy will also be judged by how well the Recordkeeper follows the rules in determining recordability of any given case. Recordability, and definitions for *"First Aid Cases"* and *"Medical Treatment Cases"* must be based on OSHA rules, and *not* on Workers' Compensation or any other *common sense* principles.
- Incidents shall be recorded within six days of knowledge that it was recordable. Questionable cases should be entered on the Log and lined out at a later date in the year if found to be not recordable. Do not wait to record a case on the Log if the case is being contested under workers' compensation. Recording a case on the OSHA Log does not bias the outcome of your contest. Section 4(b)(4) of the Occupational Safety & Health Act provides that the provisions of the Act will not affect workers' compensation liability.
- Ensure that all checks are placed in appropriate columns, and numbers are entered and tallied correctly. Body part *and type of injury* must be written at a minimum (i.e., sprain wrist, lacerate thumb). (Basic detailed instructions are printed on the back of the OSHA 200 Log).
- The right hand side of the Log from the previous year, must be tallied, signed, and posted in an employee area for the month of February. (The left side contains the names of the injured individuals. Revealing such information would violate their medical privacy rights).
- *Needlesticks of any kind, with a used needle, or needle of unknown origin, along with all "exposure incidents" as defined under the Bloodborne Pathogens Standard, are recordable as injuries on the Log. This is true, regardless if there was no time loss or no "medical treatment" provided, as in other cases.*
- *TB conversions on baseline tests, are recordable as illnesses. A positive PPD TB test conversion, from a negative baseline, for an existing employee, is recordable on the OSHA 200 Log as an illness, under column (g) and (13). New hiers with a baseline positive PPD TB test is nonrecordable.* (Depending on demographics, it is highly recommended that PPDs on new hiers be conducted twice—a booster—especially if a few years have passed since their previous test. This will do away with false negatives, due to a low immune response and will prevent the facility from appearing *"responsible"* for the conversion dur-

ing the test on the following year falsely implying that the employee may have been exposed at work. This precludes the facility from conducting epidemiological follow-up and documentation which is at best a nuisance, especially during OSHA inspections, and worst, a liability for a workers' compensation claim should the disease develop.)

C. UPDATE

In previous years, the Bureau of Labor Statistics (BLS), another agency under the Department of Labor, had collected occupational injuries and illness data from randomly selected employers each year. In 1996, the responsibility for collecting the data was transferred from the BLS to OSHA directly, and thus the solicitation and calculations of statistics will come from OSHA henceforth.

In the past, OSHA 200 Log violations have been commonly cited, but no penalties had ever been assessed. Within the past several years, deficiencies in recordkeeping have resulted, with $1600 fines on average. In two recent decisions, the Review Commission upheld the validity of OSHA citations on the Log, on an entry-by-entry basis, assessing separate counts of violations and penalties for each line item omitted or erred. Penalties in that manner have ranged from $55 to $550 each. This usually occurs only in willful and egregious cases, and pilot penalty programs are restricting recordkeeping fines to $1000.

[Note. ensure that post exposure follow-up documentation for any and all "exposure incidents" are stored and available <u>onsite</u>. This includes documentation on medical counseling, and all the necessary tests and investigations into source patients, are mandated. If employee declines testing, ensure that the declination is well documented with employee signatures. Also ensure that all privacy rights have been complied with.]

D. DETAILED SUMMARY

1. RECORDABLE CASES

Incidents are recordable on the OSHA 200 Log if they involved:

- TB conversion from a negative baseline
- *"Medical Treatment"* even without loss time (see detail below)
- Loss time (of next scheduled work day, beyond the day of the accident)
- Restricted work or motion
- *"Exposure incident"*
- Permanent damage or scarring
- Death

2. NONRECORDABLE CASES

Incidents are nonrecordable if they are *"First Aid"* cases or mere complaints, and do not fall into any of the above categories.

3. FIRST AID

"First aid treatment" cases as defined by the OSHA 200 Log Recordkeeping rules are not, by definition, recordable. First aid treatments include procedures that generally involve one-time treatment and subsequent observation of minor injuries, and do not involve loss of consciousness, restricted work, or motion or transfer to another job. Common first aid cases are

- Application of antiseptics during first visit to medical personnel
- Application of bandage(s) during any visit to medical personnel
- Use of elastic bandage(s) during first visit
- Removal of foreign bodies not embedded in the eye if only irrigation is required
- Removal of foreign bodies from wound if procedure is uncomplicated, and is, for example, by tweezers or other simple technique
- Use of nonprescription medications and administration of single dose of prescription medication on first visit for minor injury or discomfort
- Soaking therapy on initial visit or removal of bandages by soaking
- Application of hot or cold compress during first visit
- Application of ointments to abrasions to prevent drying or cracking
- Application of heat therapy during first visit
- Use of whirlpool bath therapy during first visit
- negative X-ray diagnosis
- observation of injury during visit to medical personnel

Administration of a Tetanus shot, or booster, is not, by itself, considered medical treatment. However, these shots are often given in conjunction with more serious injuries, consequently, injuries requiring these shots may be recordable for other reasons.

4. MEDICAL TREATMENT

"Medical treatment" cases as defined by the OSHA 200 Log recordkeeping rules, are, by definition, recordable, regardless if there was no loss time. The type of medical treatment that the injury necessitated dictates the recordability of the case on the Log. The medical treatment provided is the measure of determination regardless of the complaint, or nature of the incident (chemical exposure). Thus, preventative physical examinations or mere complaints, or observations at a hospital that yield no findings and no other "*medical treatment*", per se, are not recordable. It is the type of the medical treatment provided, not the nature of injury, that determines if an incident is recordable on the Log. The only exceptions are "*exposure incidents*," permanent damage, scarring, and loss of consciousness. Medical treatment cases are recordable even if they do not involve time loss. These include but are not limited to:

- Treatment of infection
- Application of antiseptics during second or subsequent visit to medical personnel
- Application of sutures (stitches)
- Treatment of second or third degree burns
- Application of butterfly adhesive dressing or steri strips in lieu of sutures
- Removal of foreign bodies embedded in the eye
- Removal of foreign bodies from the wound if the procedure is complicated because of depth of embedded foreign object, size, or location
- Use of prescription medications (except a single dose administered on first visit)
- Use of hot or cold soaking therapy during second or subsequent visit
- Application of hot or cold compress during second or subsequent visit
- Cutting away dead skin (surgical debridement)
- Application of heat therapy during second or subsequent visit
- Use of whirlpool bath therapy during second or subsequent visit
- Positive X-ray diagnosis (fractures, broken bones, etc.)
- Admission to a hospital or equivalent medical facility for treatment (not observation)
- A series of treatments given by a chiropractor is considered medical treatment

5. EXPOSURE INCIDENTS

All exposure incidents as defined under the Bloodborne Pathogens Standard are recordable, and must have appropriate medical follow-up as delineated by the Standard. This includes employee medical counseling, identifying and testing source patient if possible, offers of blood testing and blood archiving for 90 days, subsequent tests and prophylaxis for employee, etc. Follow-up documentation *must be available on site.*

6. RECORD EMPLOYEE AND WORK-RELATED INCIDENTS ONLY

Record only employee incidents, not residents, volunteers, contractors, vendors, or visitors. Record only work-related incidents. Do not confuse this with workers' compensation. That benefits law and insurance. OSHA is academic, studying the risks associated with certain job categories. For example, slip/fall in parking lots coming from and going to work, or car accidents during lunch hour while running an errand for a supervisor, may become workers' compensation cases. They are not OSHA recordable since they are not related to the job, unless the employee was sent out on a true work errand.

7. INCIDENT RECORDING

Enter a new entry for a re-aggravated old injury, only if this re-aggravation was precipitated by a *new incident/accident,* and falls within the definition of recordable injuries of course. Do not make a new entry if a flareup of an old injury occurs without a new incident. If this flareup involves additional loss time, simply increase the number of days lost already recorded on the original entry. If the initial incident was nonrecordable, but now becomes recordable because of loss time or medical treatment from the re-aggravated injury, then an entry should be made bearing the original date of accident.

8. LOSS TIME

Loss time is stricter in OSHA terms than in Workers' Compensation terms. Time lost beyond the day of the incident is considered to be lost time. Even if the incident occurred at the first minute of the shift, and the employee misses that entire shift, loss time will not be counted unless any additional days are lost beyond that. Loss time is only counted for "*scheduled*" workdays. Therefore, if an employee would have missed two days from work, but happened to be off duty those two days on his or her regular schedule, then it would not be recordable. On the other hand, if an employee is injured on Friday, and cannot work over the weekend, but was scheduled to work, then the case is recordable.

9. ACTUAL PRINTED INSTRUCTIONS ON OSHA 200 LOG FORM

The basic instructions for completing the form are printed on the back of the OSHA 200 Log Form itself. For additional guidance contact the government printing office of your local OSHA office for either of the following guides:

- "A Brief Guide to Recordkeeping Requirements for Occupational Injuries & Illnesses" (18 pages)
- "Recordkeeping Guidelines for Occupational Injuries & Illnesses" (82 pages)

E.   OSHA 101 FORM

The OSHA 101 Form is the Supplementary Record of Occupational Injury & Illness. An OSHA 101 form should be available for every entry in the Log. An in-house version can be used, as long as it contains the same information as required on the 101 Form. The information that must be contained in the report form is as follows:

- Employer name, mailing address, and location if different from mailing address.
- Employee name, social security number, home address, age, gender, occupation.
- Place of accident or exposure, whether it was on the employer's premises; what the employee was doing when injured, and how the accident occurred.
- Description of the injury or illness, including part of body affected; name of the object or substance which directly injured the employee, date of injury, or diagnosis of illness.

Appendix G

List of Agencies

OSHA—Occupational Safety and Health Administration (www.osha.gov)
Regional offices, jurisdictions, and phone numbers

Region I: Boston, MA (CT, MA, ME, NH, RI, VT)
Phone: 617-565-9860

Region II: New York, NY (NJ, NY, PR, VT)
Phone: 212-337-2378

Region III: Philadelphia, PA (DC, DE, MD, PA, VA, WV)
Phone: 215-596-1201

Region IV: Atlanta, GA (AL, FL, GA, KY, MS, NC, SC, TN)
Phone: 404-562-2300

Region V: Chicago, IL (IL, IN, MI, MN, OH, WI)
Phone: 312-353-2220

Region VI: Dallas, TX (AR, LA, NM, OK, TX)
Phone: 214-767-4731

Region VII: Kansas City, MO (IA, KS, MO, NE)
Phone: 816-426-5861

Region VIII: Denver, CO (CO, MT, ND, SD, UT, WY)
Phone: 303-844-1600

Region IX: San Francisco, CA (American Samoa, AZ, CA, Guam, HI, NV)
Phone: 415-975-4310

Region X: Seattle, WA (AK, ID, OR, WA)
Phone: 206-553-5930

EPA—Environmental Protection Agency (www.epa.gov)
Regional offices, jurisdiction, and asbestos coordinator phone numbers

Region I: (CT, ME, MA, NH, RI, VT)
Phone: 617-565-3835

Region II: (NJ, NY, PR, VI)
Phone: 201-321-6671

Region III: (DE, DC, MD, PA, VA, WV)
Phone: 215-597-3160

Region IV: (AL, FL, GA, KY, MS, NC, SC, TN)
Phone: 404-347-5014

Region V: (IL, IN, MI, MN, OH, WI)
Phone: 312-886-6003

Region VI: (AR, LA, NM, OK, TX)
Phone: 214-655-7244

Region VII: (IA, KS, MO, NE)
Phone: 913-551-7020

Region VIII: (CO, MT, ND, SD, UT, WY)
Phone: 303-293-1442

Region IX: (AZ, CA, HI, NV, American Samoa, Guam)
Phone: 415-556-5406

Region X: (AK, ID, OR, WA)
Phone: 206-442-4762

GPO—Government Printing Office
Phone: 202-512-0132/2356
www.gpo.gov

NIBS—National Institute of Building Science, Washington, D.C.
Phone: 202-289-7800

NIOSH—National Institute of Occupational Safety & Health
Phone: 1-800-35-NIOSH
www.dcd.gov/niosh

NIST—National Institute for Standards and Technology (contact for lab accreditation)
Gaithersburg, MD
Phone: 301-975-4016

CDC—Centers for Disease Control (www.cdc.gov)
ATSDR Agency for Toxic Substances and Disease Registry
Phone: 888-422-8737
CDC HIV Registry for Healthcare Workers
Phone: 888-737-4448

Index

.

x